PALEOECOLOGY

PALEOECOLOGY
Past, Present, and Future

DAVID J. BOTTJER

WILEY

Registered office: John Wiley & Sons, Ltd, The Atrium, Southern Gate, Chichester, West Sussex, PO19 8SQ, UK

Editorial offices: 9600 Garsington Road, Oxford, OX4 2DQ, UK

The Atrium, Southern Gate, Chichester, West Sussex, PO19 8SQ, UK

111 River Street, Hoboken, NJ 07030-5774, USA

For details of our global editorial offices, for customer services and for information about how to apply for permission to reuse the copyright material in this book please see our website at www.wiley.com/wiley-blackwell.

Library of Congress Cataloging-in-Publication Data

Bottjer, David J.
  Paleoecology : past, present, and future / David J. Bottjer.
      pages cm
  Includes bibliographical references and index.
  ISBN 978-1-118-45586-9 (cloth) – ISBN 978-1-118-45584-5 (pbk.) 1. Paleoecology. 2. Ecology. 3. Global environmental change. I. Title.
  QE720.B66 2016
  560′.45 – dc23

                                                                                                        2015034607

A catalogue record for this book is available from the British Library.

Wiley also publishes its books in a variety of electronic formats. Some content that appears in print may not be available in electronic books.

Cover image should be "©Aneese/istockphoto"

Set in 10/12pt MinionPro by SPi Global Private Limited, Chennai, India

1   2016

Contents

Color plate pages fall between pp. 1 and 42

Preface

This book is intended for advanced undergraduates and beginning graduate students who will have had an undergraduate course in paleontology as geology or earth science majors or a class in ecology and evolution as biology majors. It is also aimed at professionals who want to discover what modern paleoecology with an evolutionary and conservation paleoecology emphasis looks like. It is not aimed to be encyclopedic in nature but rather as an introduction to many of the fascinating aspects of paleoecology. The approach has been to broadly cover paleoecology, but the focus is deep-time marine paleoecology, as that is where my experience lies. Paleoecology has typically been focused on the past, but its relevance to managing ecosystems in the future has become more and more apparent, and it is hoped that this text will stimulate further research in this fashion.

The structure of this book is to present an easy-to-read text, with more details in the figures and figure captions. Thus, the text is meant to provide a broad overview, while the figures and figure captions provide added depth. With this approach, my hope is that readers won't get bogged down in a detailed text, but can find those details in the figures and captions.

Development of this book has been the product of my interactions with many people. I thank my undergraduate mentor Bruce Saunders and my Ph.D. advisor Don Hattin, as well as other graduate mentors Gary Lane, Bob Dodd, Dick Beerbower, Paul Enos, and Don Kissling. At USC, I have been stimulated on a daily basis by colleagues Bob Douglas, Al Fischer, Donn Gorsline, Frank Corsetti, Will Berelson, and Josh West. My collaborations with those from other institutions including Bill Ausich, David Jablonski, Luis Chiappe, Eric Davidson, Bill Schopf, and Junyuan Chen have been inordinately fruitful. But my major collaborators over the years have been my graduate students, and I especially thank Chuck Savrda, Mary Droser, Jennifer Schubert, Kate Whidden, Kathy Campbell, Carol Tang, Reese Barrick, James Hagadorn, Adam Woods, Steve Schellenberg, Nicole Fraser, Nicole Bonuso, Sara Pruss, Steve Dornbos, Margaret Fraiser, Pedro Marenco, Katherine Marenco, Catherine Powers, Scott Mata, Rowan Martindale, Kathleen Ritterbush, Lydia Tackett, Carlie Pietsch, Liz Petsios, Jeff Thompson, and Joyce Yager. I am indebted to Patricia Kelley and Paul Taylor who provided thorough reviews of this book in manuscript form and Ian Francis and Kelvin Matthews of Wiley-Blackwell who have provided much encouragement and assistance in the publication process. My parents John and Marilyn Bottjer have supported and encouraged me through all these years. My wife Sarah Bottjer has been the essential person enabling me to pursue a life focused on paleoecology and paleobiology.

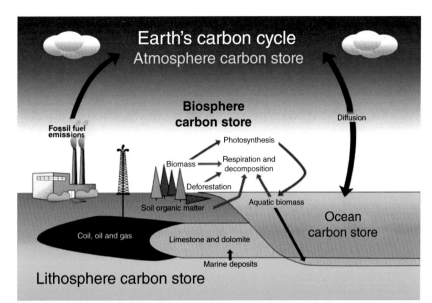

Figure 1.4 Schematic of modern carbon cycle including anthropogenic influence. From the New York State Department of Environmental Conservation website: http://www.dec.ny.gov/energy/76572.html.

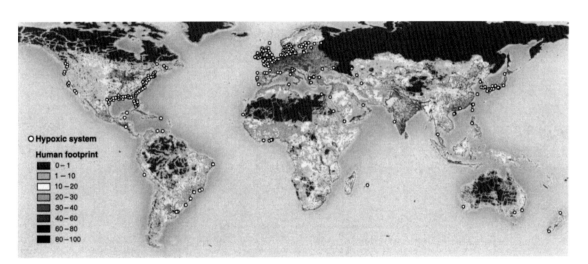

Figure 1.6 Location of hypoxic system coastal "dead zones." Their distribution matches the global human footprint, where the normalized human influence is expressed as a percent, in the Northern Hemisphere. For the Southern Hemisphere, the occurrence of dead zones is only recently being reported. From Diaz and Rosenberg (2008). Reproduced with permission from the American Association for the Advancement of Science.

Paleoecology: Past, Present and Future, First Edition. David J. Bottjer.
© 2016 John Wiley & Sons, Ltd. Published 2016 by John Wiley & Sons, Ltd.

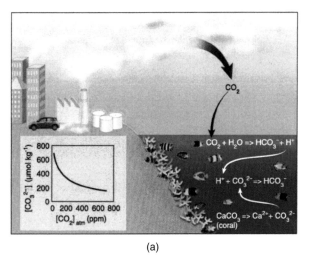
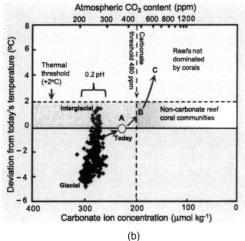

(a)   (b)

Figure 1.7 Increase in atmospheric carbon dioxide and its influence on ocean acidification and the resultant affect on development of coral reefs in the past, present and future. (a) Increased carbon dioxide concentration in the oceans leads to decreased availability of carbonate ions, which are needed by corals to secrete their skeletons made of calcium carbonate. (b) Plot of temperature, atmospheric carbon dioxide content, and ocean carbonate ion concentration showing the predicted trend in the future of reefs not dominated by corals with increased levels of acidification. From Hoegh-Gulberg et al. (2007). Reproduced with permission from the American Association for the Advancement of Science.

Figure 1.9 The net radiative forcing due to changes in atmospheric carbon dioxide concentration and total solar irradiance from 5 to 45 million years ago. The three curves represent the range in carbon dioxide concentration using three different proxies for ancient atmospheric carbon dioxide. The shaded area denotes the range in radiative forcing projected to occur by 2100 according to the Intergovernmental Panel on Climate Change (IPCC) Fourth Assessment Report (2007). Net radiative forcing is in watts per square meter. From Kiehl (2011). Reproduced with permission from the American Association for the Advancement of Science.

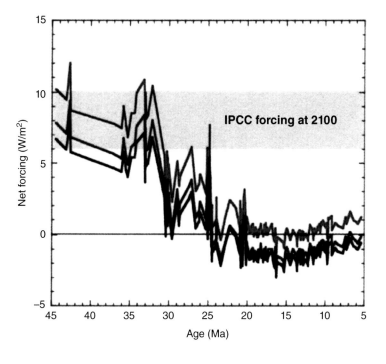

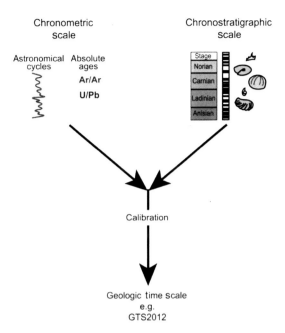

Figure 2.1 Steps in construction of a geological time scale. The chronostratigraphic scale is a relative time scale and includes (from left to right) formalized definitions of geologic stages (here with examples of Triassic stages), magnetic polarity zones, and biostratigraphic zonation units, with examples here indicated by fossil symbols (from top to bottom, conodont, ammonoid, echinoderm, foraminifer, bivalve). The chronometric scale is measured in years and includes absolute ages measured from radiogenic isotope systems such as Argon/Argon and Uranium/Lead, and astronomical cycles exemplified by the sedimentary expression of Earth's orbital cycles. If some or all of this chronometric and chronostratigrahic information is available it can then be merged to produce an age calibration that allows linkage into a formal geologic time scale, here indicated as that found in Gradstein et al. (2012). From Gradstein et al. (2012). Reproduced with permission from Elsevier.

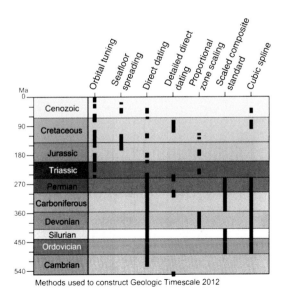

Methods used to construct Geologic Timescale 2012

Figure 2.2 Methods used to construct the geological time scale for the Phanerozoic in Gradstein et al. (2012), which depend on the quality of data available for different time intervals. From Gradstein et al. (2012). Reproduced with permission from Elsevier.

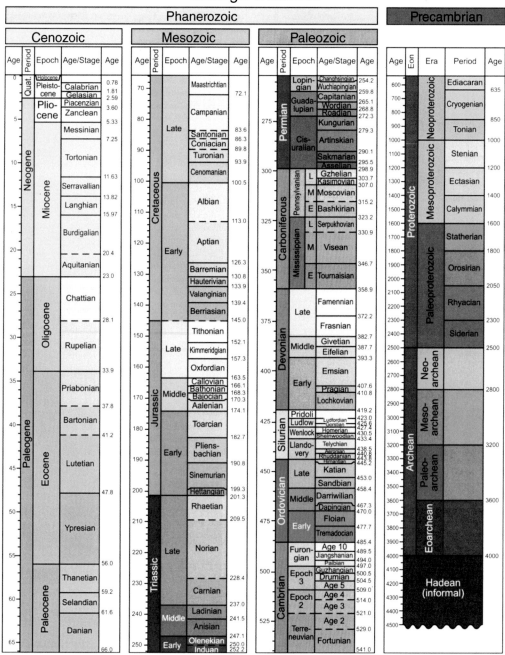

Figure 2.3 Geological time scale from Gradstein et al. (2012). Reproduced with permission from Elsevier.

Figure 2.6 Reconstruction of a Carboniferous forest including a dragonfly with a wingspan of 60 cm. From Kump et al. (2009).

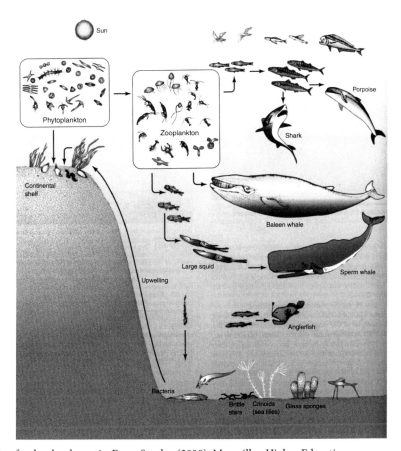

Figure 3.2 Marine food web schematic. From Stanley (2008). Macmillan Higher Education.

Figure 3.4 Model of the Late Cretaceous pterosaur *Quetzalcoatlus northropi* in flight. This halfscale working model was designed and built by a team at Aerovironment, Inc., led by Paul MacCready, an aeronautical engineer. When alive this pterosaur had a wingspan of 11 m. AeroVironment, Inc. website: http://www.avinc.com/uas/adc/quetzalcoatlus/. Reproduced with permission.

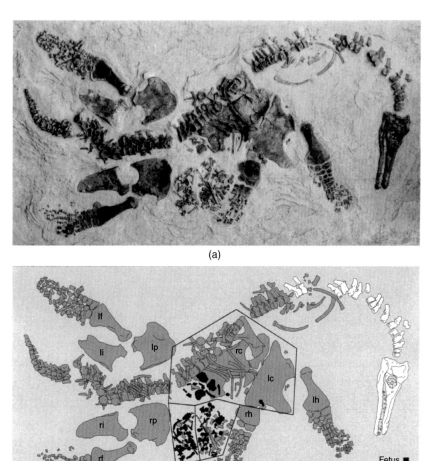

(a)

(b)

Figure 3.5 Skeleton (a) and interpretive drawing (b) of a fetus preserved within a pregnant adult Late Cretaceous plesiosaur *Polycotylus latippinus*. This is the first definitive evidence that plesiosaurs were viviparous. This evidence includes skeletal features showing that the smaller individual is a juvenile, taphonomic evidence that the juvenile was not consumed by the adult, articulation features of the juvenile skeleton indicating that it was within the adult at the time of burial, and skeletal features showing that both skeletons are *P. latippinus*. From O'Keefe and Chiappe (2011). Reproduced with permission from the American Association for the Advancement of Science.

Figure 3.6 Specimen of the Early Cretaceous bird *Yanornis* (a) with ingested whole fish in the crop (b) and macerated fish bones in the ventriculus (c). Scale bars are 5 cm for (a), 1 cm for (b) and 1 cm for (c). From Zheng et al. (2014). Used under CC-BY-3.0 https://creativecommons.org/license/by/3.0.

Figure 4.5 A Permian assemblage exhibiting parts of a variety of plants. This preserves an autochthonous Late Permian forest-floor litter in which leaves of the gymnosperm *Glossopteris* (darkest brown fossil at top of lower right quadrant) are preserved with horsetail (sphenopsid) groundcover which includes *Trizygia* (brown leaves in upper center and upper right quadrant; a yellow cone in left-center of lower left quadrant) and *Phyllotheca* (two foliar whorls showing long, narrow leaves that fit into a funnel-shaped sheath, in left-center of upper left quadrant). Such litter horizons, from the Upper Permian Balfour Formation, are an uncommon feature of the sequence transitioning the Permian–Triassic boundary, as defined by vertebrate biostratigraphy, in the Karoo Basin, South Africa. Locality and additional paleoenvironmental information can be found in Prevec et al. (2010) and Gastaldo et al. (2014). Photograph by Robert A. Gastaldo. Reproduced with permission.

Figure 4.6 Silicification of fossil forests buried by volcanic eruptions. On the western side of the island of Lesvos, Greece, volcanogenic debris flows and lahars preserve autochthonous Miocene forests, where erect, silicified trunks of *Taxodioxylon* (*Sequoia*) are preserved to a height of 7.1 m. The 350 m-thick Miocene volcaniclastic sequence, known as the Sigri Pyroclastic Formation, erupted from the Varossa stratovolcano, and buried at least two major forest horizons. From Mpali Alonia Park in the Lesvos Petrified Forest GeoPark, Robert A. Gastaldo for scale. Photograph by Robert A. Gastaldo. Reproduced with permission.

Figure 4.10 Crinoid assemblage from the Lower Mississippian Maynes Creek Formation, Legrand, Iowa. Width of the darkest, common crinoid is 1 cm. Photograph by William I. Ausich. Reproduced with permission.

Figure 4.14 Ediacara biota fossils on the "E" surface from the Mistaken Point Formation, showing the stalked suspension feeder *Charniodiscus spinosus* (disc with attached frond below coin), and various rangeomorphs. Coin is 24 mm. Additional information can be found in Narbonne et al. (2007). Photograph by David J. Bottjer.

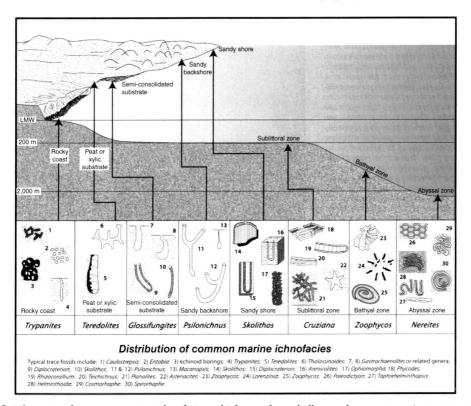

Figure 5.3 Schematic of common marine ichnofacies, which vary from shallow to deep-water environments and substrate type. Sketches of typical trace fossils for each ichnofacies are shown; the name of each ichnofacies (e.g., *Trypanites*) reflects a characteristic trace fossil for that ichnofacies. To identify a particular ichnofacies the trace fossil which gives that ichnofacies its name does not need to be present. From MacEachern et al. (2010). Reproduced with permission from the Geological Association of Canada.

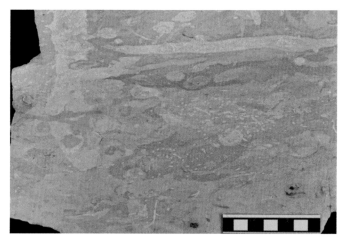

Figure 5.5 Thorough bioturbation resulting in ichnofabric index (ii) 5 developed at a marly chalk–chalk transition (mid- to outer shelf) in the Upper Cretaceous (Campanian) Demopolis Chalk (Selma Group) exposed in western Alabama (eastern Gulf coastal plain), USA. Distinct lighter-filled trace fossils include *Thalassinoides* (wide vertical structure on left side and probably highest long horizontal structure near top), *Zoophycos* (horizontal structure with faint meniscate backfill below left side of horizontal *Thalassinoides*?), and *Chondrites* (lightest structures including dots to short branching segments). Additional information can be found in Locklair and Savrda (1998). Scale is in centimeters. Photograph by Charles E. Savrda. Reproduced with permission.

Figure 5.11 Two parallel sauropod trackways from the Upper Cretaceous of Bolivia. The direction of progression is down dip, towards the bottom of the photograph. There also is a theropod trackway oriented from right to left and crossing the lower third of the sauropod trackways. These are from the Cal Orcko Quarry site on the outskirts of Sucre, Bolivia. They occur on a limestone bedding plane that is part of a mosaic of lacustrine and marginal lacustrine facies of the Upper Cretacoues El Molino Formation (Maastrichtian). Additional information can be found in Lockley et al. (2002). Photograph by Martin Lockley. Reproduced with permission.

Figure 6.2 A group of columnar stromatolites that are part of a bioherm preserved in dolostone. Individual stromatolites are on the order of tens of centimeters tall, and the bioherm extends laterally for several meters. From the Ediacaran Deep Spring Formation at Mt. Dunfee, Nevada, United States. Additional information can be found in Oliver and Rowland (2002). Photograph by Russell S. Shapiro. Reproduced with permission.

Figure 6.5 Conical stromatolites from the Belt Supergroup (ca. 1.4 billion years old) in southern Alberta, Canada. Photograph by Frank A. Corsetti. Reproduced with permission.

Figure 6.7 Bedding plane view of typical Neoproterozoic tube-forming stromatolites from the Noonday Dolomite at Winters Pass Hills, in the Death Valley area, eastern California, United States. Tubes are filled with dark micrite/microspar. Additional information can be found in Corsetti and Grotzinger (2005). Photograph by Frank A. Corsetti. Reproduced with permission.

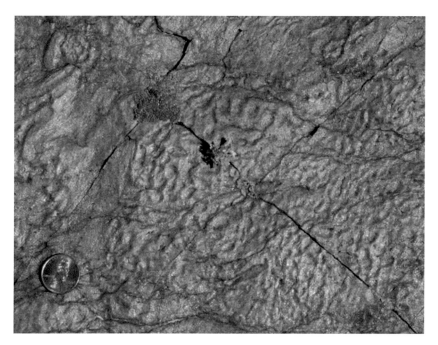

Figure 6.8 Bedding plane surface of the lower Cambrian Harkless Formation with quasi-polygonal, moderately flat-topped wrinkle structures. Locality is near Cedar Flats, White-Inyo Mountains, CA, United States. Additional information can be found in Hagadorn and Bottjer (1997, 1999) and Bailey et al. (2006). Photograph by Stephen Q. Dornbos. Reproduced with permission.

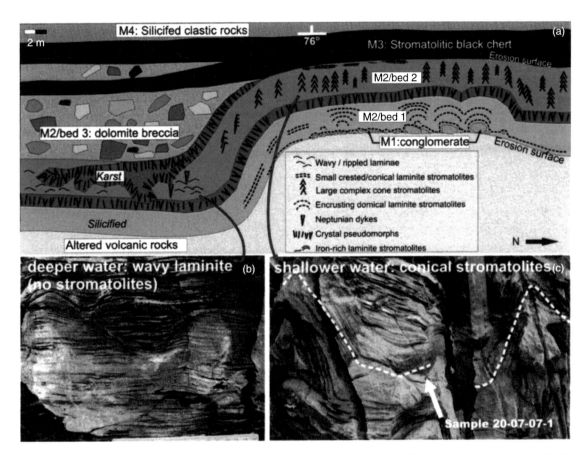

Figure 7.1 Archean (~3.45 billion years old) reef-like assemblage of stromatolites from the Strelley Pool Formation at the platform margin outcrop on southern Trendall Ridge (Western Australia). (a) Outcrop map showing cross-sectional view of stratigraphy from underlying altered volcanic rocks up through members (M) 1–3 and part of member 4 of the Strelley Pool Formation. Of special interest is the paleotopographic feature, with stromatolites only deposited on the high side (right). (b) Wavy laminites deposited in deeper water south (left) of the paleohigh. (c) Large complex conical stromatolites formed on the paleohigh. The dotted white line traces a single lamina across two coniform stromatolites. Scale rule in b and c is 15 cm. From Allwood et al. (2009). Reproduced with permission from the National Academy of Sciences.

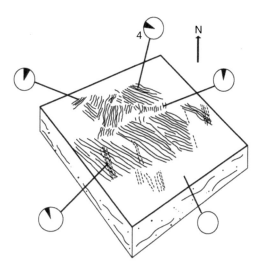

Figure 7.2 Archean (2.9 billion years old) microbially induced sedimentary structures (MISS) from the Nhlazatse Section, Pongala Supergroup, South Africa. These multidirected ripple marks which were deposited in tidal flat settings are exposed on a 7 to 10 cm thick fine sandstone bed; scale: 1 m. The ripple pattern is composed of patches (or groups) of ripple marks. Each patch (or group) is characterized by one ripple crest direction as shown in the stereonets. The numbers 1–4 indicate the order of formation of the ripple generations. The first generation ripple marks are least visible, and mostly occupy the smallest area on the bedding plane: (1) 5%; (2) about 1.5%; (3) 17%; (4) 45%; (nonpreserved) 33%. From Noffke et al. (2008). Reproduced with permission from John Wiley & Sons.

Figure 7.5 Diorama depicting the Ediacara Biota from South Australia, including a large *Dickinsonia* (center top), *Spriggina* (lower right corner), *Kimberella* (lower left center), and *Tribrachidium* (lower left corner), as well as a variety of fronds. From Zimmer (2006).

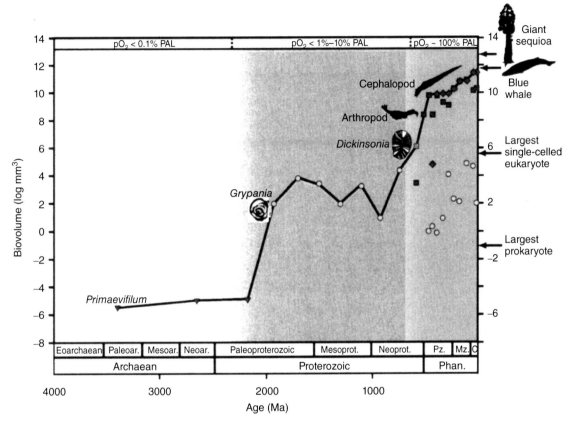

Figure 7.9 Change in size of largest fossils through Earth history. Size maxima are plotted as computed biovolume (log mm³). Single-celled eukaryotes, animals, and vascular plants are illustrated separately for the Ediacaran and Phanerozoic. The solid line denotes the trend in the overall maximum for all of life. Increases in the overall maximum occurred in discrete steps approximately corresponding to increases in atmospheric oxygen levels in the mid-Paleoproterozoic and Ediacaran-Cambrian-early Ordovician. Sizes of the largest fossil prokaryotes were not compiled past 1.0 Gya. Estimates of oxygen levels are expressed in percentages of the present atmospheric level (PAL). Phan., Phanerozoic; Pz, Paleozoic; Mz, Mesozoic; C, Cenozoic. Red triangles, prokaryotes; yellow circles, protists; blue squares, animals; black diamonds, vascular plants; gray square, probable multicellular eukaryote. From Payne et al. (2009). Reproduced with permission from the National Academy of Sciences.

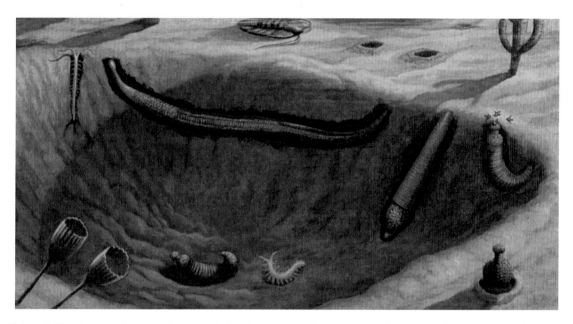

Figure 7.15 Schematic diagram of Burgess Shale infauna envisaged as if a modern submersible was there and had scooped a sample of seafloor sediment. The infauna is dominated by priapulid worms, of which the most abundant was *Ottoia*. In this scene three individuals are visible; one on the floor of the large excavation, another in the process of consuming hyoliths (mid-right), while the third is emerging from its burrow and displaying its spinose proboscis (lower right). Two other priapulids are visible in the excavation: the elongate, more-or-less horizontal worm is *Louisella*, shown here in its life position as a sedentary animal occupying an elongate burrow with openings to the overlying seawater at either end. The animal inclined downwards, with its posterior end just emerging from the seafloor, is an example of *Selkirkia*. It inhabited a parchment-like tube, and in common with other priapulids had a spiny proboscis that was employed, when necessary, for burrowing. The other type of worm visible in the excavation are two examples of the polychaete annelid *Burgessochaeta*, with one individual wriggling on the floor and the other in its burrow with anterior tentacles extending sideways (far left). Above the seafloor are a typical Burgess Shale sessile epifauna – a trilobite (top), the sponge *Vauxia* (upper right), and the enigmatic *Dinomischus* (lower left). From Conway Morris (1998). Reproduced with permission from Oxford University Press.

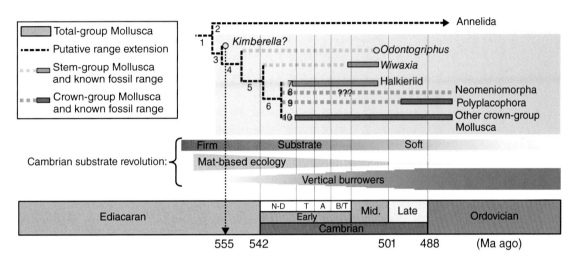

Figure 7.16 Evolution of the molluscs and the Cambrian substrate revolution. A proposed evolutionary tree for molluscs indicates this sequence of evolution of various characters during the Ediacaran and Cambrian. 1: Protostome bilaterians; serial replication; triploblastic. 2: Segmentation by coelomic metameres. 3: Large size; with iteration but not coelomic segmentation; ovoid; dorso-ventrally flattened; stiffened cuticular dorsum; flat, noncuticularized ventral sole; radula of iterated, paired mirror-image teeth and radular membrane (certain for *Odontogriphus*); feeding on biomat? 4: Groove (mantle cavity) between dorsum and ventrum with serial ctenidia; paired salivary glands; straight digestive tract; nervous system ladder-like?; coelom posterior; restricted to reproductive and excretory organs? 5: Non-calcified scleritome, sclerites arranged in three mirror-image longitudinal zones. 6: Calcification of epidermally nucleated sclerites that pass through cuticle; calcified shell from serial shell fields; no periostracum from periostracal groove of mantle lobe. 7: Two shell fields. 8: Tubiform; reduced foot; sclerites in 1–3 longitudinal rows beside foot groove; progenetic loss of gills and shells; embryological evidence of vestigial shell fields. 9: Eight or more shell fields; sclerites not in longitudinal zones. 10: Loss of sclerites and serial shell fields; true periostracum secreted from mantle lobe; shells paired or single, reduction of gills; further variety of body plans. A, Atdabanian; B/T, Botomian/Toyonian; N-D, Nemakit-Daldynian; T, Tommotian. From Caron et al. (2006). Reproduced with permission from Macmillan Publishers Ltd.

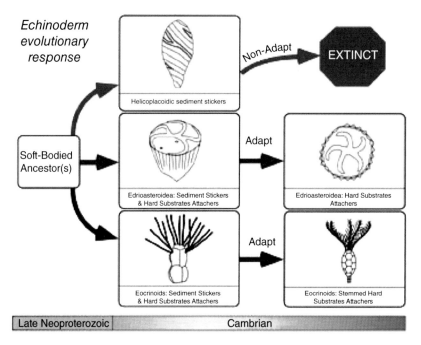

Echinoderm evolutionary response

Helicoplacoidic sediment stickers

Non-Adapt

EXTINCT

Soft-Bodied Ancestor(s)

Adapt

Edrioasteroidea: Sediment Stickers & Hard Substrates Attachers

Edrioasteroidea: Hard Substrates Attachers

Adapt

Eocrinoids: Sediment Stickers & Hard Substrates Attachers

Eocrinoids: Stemmed Hard Substrates Attachers

Late Neoproterozoic   Cambrian

Figure 7.17 Evolution of the echinoderms and the Cambrian substrate revolution. Arrows do not represent a direct evolutionary relationship between specific echinoderms shown, but imply a general evolutionary trend through the Cambrian within each of the groups examined, with these echinoderms serving as individual examples. Helicoplacoid drawing is from specimen 3 cm in height. For edrioasteroids, *Camptostroma* (left) drawing is from specimen 5 cm in height. Edrioasteroid on right is a schematic of a typical attaching edrioasteroid with width size of 5 cm. For eocrinoids, *Lichenoides* (left) drawing is from a specimen 2.5 cm in height. Eocrinoid on right is *Tatonkacystis*, drawing is from specimen approximately 5 cm in height. Geological time not to scale and boxes do not represent the precise age range of the echinoderms they contain. From Bottjer et al. (2000). Reproduced with permission from the Geological Society of America.

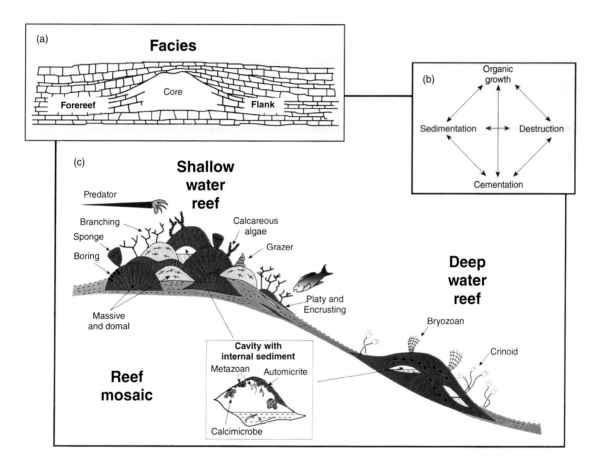

Figure 9.1 Development of reefs in modern and ancient environments. Schematic diagrams illustrating (a) cross-sectional geometry of a typical reef as exposed in outcrop, (b) complex relationships between biological and sedimentological processes that control reef composition, and (c) ecological and sedimentological attributes of shallow-water and deep-water reefs. From James and Wood (2010). Reproduced with permission from the Geological Association of Canada.

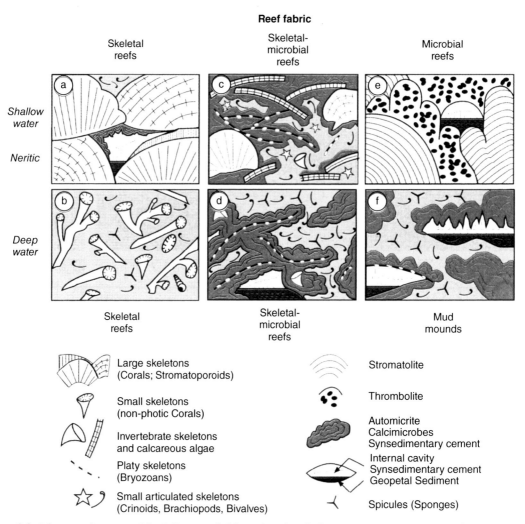

Figure 9.3 Schematic diagrams of the different rock fabrics found in shallow-water and deep-water reefs, with typical biotic and abiotic components. From James and Wood (2010). Reproduced with permission from the Geological Association of Canada.

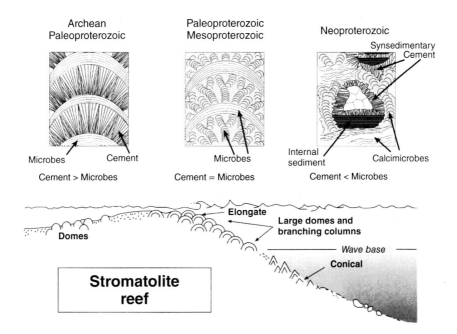

Figure 9.4 Precambrian stromatolite reefs. Schematic diagrams show Archean–Neoproterozoic reef fabrics of cement and microbial features (upper), and a zoned Proterozoic marginal reef (lower). In the Archean–Paleoproterozoic the amount of synsedimentary cement was generally greater than the microbial content; in the later part of the Paleoproterozoic and Mesoproterozoic they were about equal; in the Neoproterozoic microbes and calcimicrobes were volumetrically more abundant than synsedimentary cement. Note characteristic trends in stromatolite morphology across reef environments. From James and Wood (2010). Reproduced with permission from the Geological Association of Canada.

Figure 9.5 Branching archaeocyath sponge and surrounding sediment, including archaeocyath fragments. This image is from an outcrop of the Poleta Formation in Esmeralda County, Nevada, USA. Branches of archaeocyath are ~5 mm wide. Additional information can be found in Rowland (1984). Photograph by Stephen M. Rowland. Reproduced with permission.

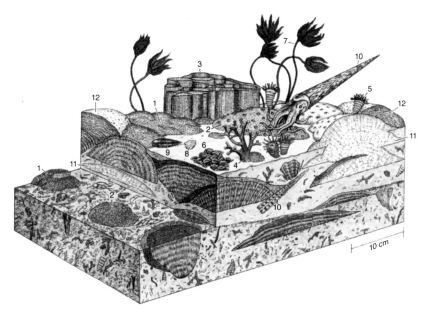

Figure 9.7 Reconstruction of a Silurian reef. This reef is composed mostly of stromatoporoids and tabulate corals. 1: tabulate coral (*Favosites*); 2: tabulate coral (*Heliolites*); 3: tabulate coral (*Halysites*); 4: bryozoan; 5: rugose coral; 6: spiriferid brachiopod; 7: crinoid; 8: brachiopod; 9: trilobite; 10: orthocone nautiloid; 11: stromatoporoid; 12: thrombolite (Copyright John Sibbick). From James and Wood (2006). Reproduced with permission from the Geological Association of Canada.

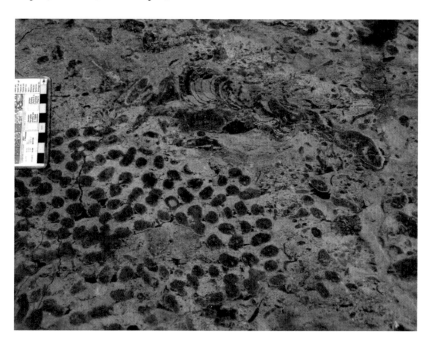

Figure 9.8 Cut slab of the upper Rhaetian (Late Triassic) Adnet reef from the Tropfbruch Quarry (Oberrhätriffkalk Foormation) near Adnet, Salzburg, Austria. Filling the lower left quadrant is a cross section through a colony of *Retiophyllia* sp. phaceloid corals (subcircular, dark grey spots). Visible as the largest fossil in the upper right hand quadrant is a sphinctozoan sponge. These are surrounded by reef sediment including solitary corals. Scale is in centimeters. Additional information can be found in Bernecker et al. (1999). Photograph by Rowan C. Martindale. Reproduced with permission.

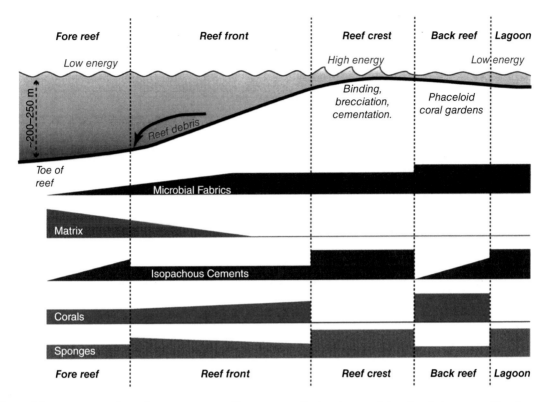

Figure 9.9 Compositional trends in an enormous Upper Triassic reef. This is an idealized environmental transect (top) 5 km long through the Rhaetian (Upper Triassic) Gosausee reef (Gosau, Austria), one of the Dachstein reefs, showing trends in microfacies composition for different reef facies (bottom). From Martindale et al. (2013). Reproduced with permission from Elsevier.

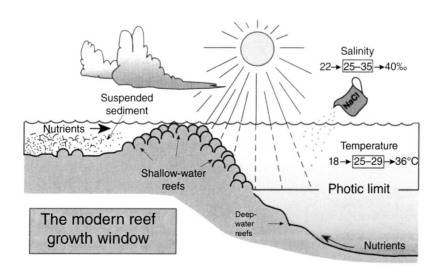

Figure 9.11 Schematic of the environmental parameters affecting the growth of modern coral reefs. Although reefs are typically low nutrient systems, nutrients can be supplied to shallow reefs from run-of, including suspended sediment, and to deep water reefs through upwelling. Other important components include temperature, salinity, and the availablility of sunlight. From James and Wood (2010). Reproduced with permission from the Geological Association of Canada.

Figure 9.13 Bedding plane view of a Middle Triassic shell bed composed primarily of the terebratulid brachiopod *Aulacothyroides liardensis*. This shell bed occurs within the Liard Formation outcropping along the shore of Williston Lake, British Columbia, Canada. Knife is 3.5 cm long. Additional information can be found in Greene et al. (2011). Photograph by David J. Bottjer.

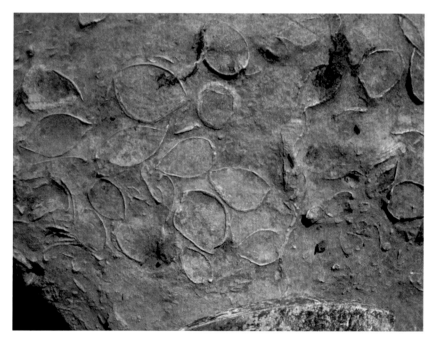

Figure 9.17 Cross-sections of articulated lucinid bivalves from the Lower Miocene Ugly Hill seep deposit. This site occurs in the uplifted East Coast Basin forearc in the Tolaga Group on the eastern North Island of New Zealand. Head of rock hammer for scale. Additional information can be found in Saether et al. (2010). Photograph by Kathleen A. Campbell. Reproduced with permission.

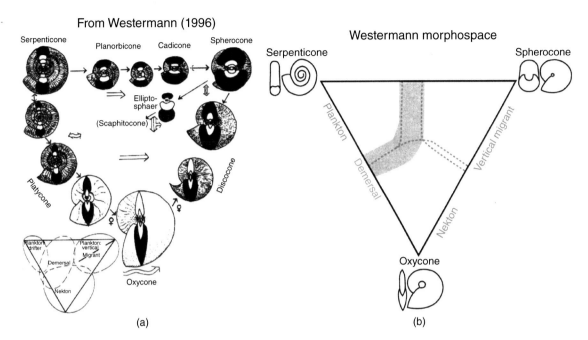

Figure 10.7 Paleoecology of planispiral ammonoids. This is portrayed as Westermann morphospace, where planispiral ammonoids of different shapes are associated with different hypothetical life modes, as synthesized by Westermann (1996). (a) Summary of ammonoid mobility for common planispiral shell shapes that grade between three forms – serpenticone, sphaerocone, and oxycone. Hypothetical life modes are indicated by the inset triangle. Each illustration includes a side view of the outer shell and a cross-section through the whorls, superimposed. (b) Westermann Morphospace, based on (a). The three components of shell shape (exposure of the umbilicus, overall inflation, and whorl expansion) for ammonite specimens dictate their placement within this ternary diagram. Serpenticones (high umbilical exposure, low overall inflation, low whorl expansion) were incapable of directed swimming and plot in the plankton field. Sphaerocones (low umbilical exposure, high inflation, and low whorl expansion) are interpreted to have moved up and down in the water column and thus plot in the vertical migrant field. Oxycones (low umbilical exposure, low inflation, and high whorl expansion) would have been efficient swimmers and thus plot in the nekton field. Platycones and planorbicones were mobile, moving along the seafloor, and thus plot in the demersal field. Details on terminology of ammonoid morphology can be found in Ritterbush and Bottjer (2012). From Ritterbush and Bottjer (2012). Reproduced with permission from the Cambridge University Press.

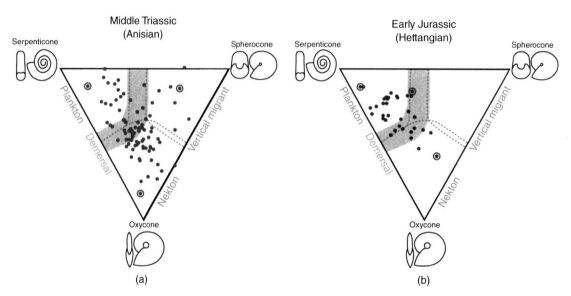

Figure 10.8 Comparison of ammonoid shell shapes and hypothetical life modes in Westermann Morphospace. (a) Plot of Middle Triassic ammonoids of Nevada include each major morphotype. Each point represents the largest measurable specimen of each species ($N = 85$) in the collections. (b) Earliest Jurassic ammonoids of Nevada include a comparatively limited variety of shell shapes. Each point represents the largest measurable specimens of each species ($n = 35$). All specimens were measured from monographs with calipers. Measurement error is about 1.7 mm. When repeat measurements of a specimen are plotted together in Westermann Morphospace, the plotted points partially overlap; the error is not great enough to cause noticeable change in the position of points. Specimens corresponding to circled data points are illustrated in Ritterbush and Bottjer (2012). From Ritterbush and Bottjer (2012). Reproduced with permission from the Cambridge University Press.

Early amphibian

Tiktaalik

Lobe-finned fish

Figure 11.2 Vertebrates evolve to inhabit terrestrial ecosystems. This shows the habitat and evolution of the Late Devonian sarcopterygian fish *Tiktaalik*. On left the skeleton is shown below, with a reconstruction of *Tiktaalik* shown above in a shallow freshwater environment marginal to land where it lived. *Tiktaalik* was intermediate between lobe-finned fishes and amphibians and had a flat, crocodile-like skull with eyes and nostrils on the upper surface. Its fins were intermediate in form between those of lobe-finned fishes and amphibians, with shoulder joints and fingerlike bones which would have permitted it to prop up the front of its body. In shallow water this would have allowed it to survey its surroundings. Three bones in its front fin are portrayed on the right in color which show homologies with bones of lobe-finned fishes and early amphibians. *Tiktaalik* ranged from 2 to 3 m in length. Illustration by Zina Deretsky, photo by Ted Daeschler. From Stanley (2008).

Figure 11.3 Upright stump of the lycopsid *Sigillaria* rooted into the top of a coal seam (Herrin Coal) which is Middle Pennsylvanian (Desmoinesian) in age. Tree stump, located at the end of the extended tape measure, is encapsulated in tidal rhythmites (interlaminated silt and silty mud), and is preserved as described in Fig. 11.4. White colored "dust" on coal mine walls and ceiling is limestone dust used for coal dust suppression. Man at left (William A. DiMichele) for scale. Photograph by Scott D. Elrick. Reproduced with permission.

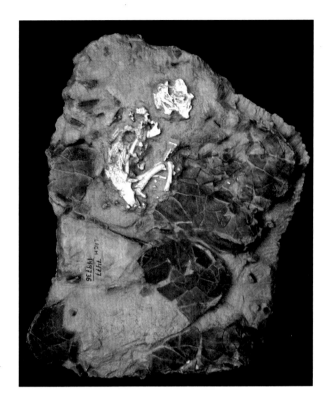

Figure 11.6 Eggs (black) and neonate (white) of *Gigantoraptor erlianensis*, an oviraptorid theropod dinosaur. This specimen (LACMA 7477/149736) is from Sanlimiao in Henan Province, China, and was found in the Upper Cretaceous (Maastrichtian) Zoumagang Formation deposited in the Xixia Basin. Embryo is 25 cm from top to bottom. Photograph by Luis Chiappe. Reproduced with permission.

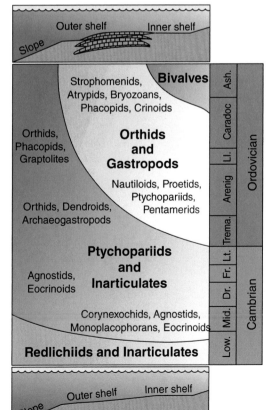

Figure 12.3 Onshore–offshore occurrence of assemblages during the Cambrian and Ordovician. Taxa in larger bold lettering are those which were most significant in differentiating the assemblages through cluster analysis (Sepkoski & Sheehan, 1983). The lower two groups of assemblages correspond to the Cambrian Evolutionary Fauna, the third group of assemblages corresponds to the Paleozoic Evolutionary Fauna, and the assemblages dominated by bivalves correspond to the Modern Evolutionary Fauna (see Fig. 3.11). Trilobites include redlichiids, corynexochids, agnostids, ptychopariids, phacopids and proetids; brachiopods include inarticulates, orthids, atrypids and strophomenids. Below, the classic onshore–offshore environmental gradient, above, the same gradient modified with the addition of midshelf carbonate buildups. From Harper (2006). Reproduced with permission from Elsevier.

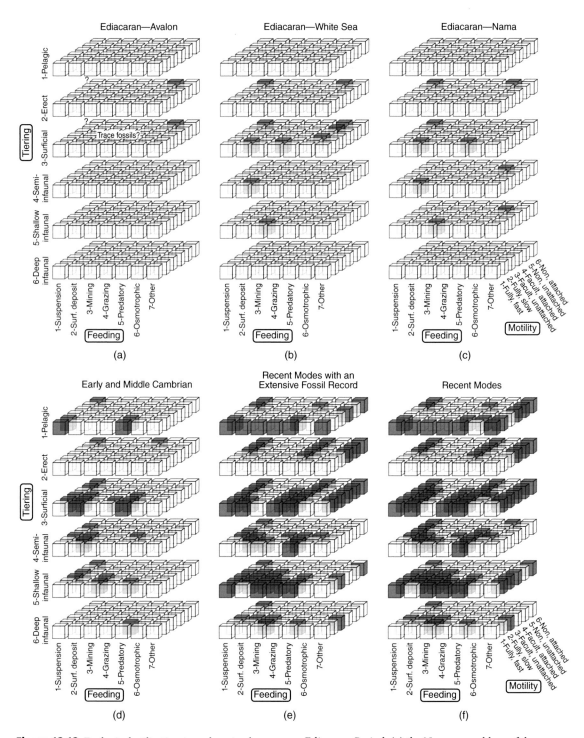

Figure 12.12 Ecological utilization in early animal faunas and the modern marine fauna using the Bush cube (Fig. 3.13). Colored boxes denote modes of life occupied by at least one species in a fauna, and white boxes denote unoccupied modes. Gray boxes and question marks in (a) indicate uncertainty. (a) The Avalon assemblage of the Ediacaran Period, (b) the White Sea assemblage of the Ediacaran Period, (c) the Nama assemblage of the Ediacaran Period, (d) the early-Middle Cambrian Period, (e) Recent animals that have an extensive fossil record, and (f) all Recent animals. From Bush et al. (2011). Reproduced with permission from Springer Science and Business Media.

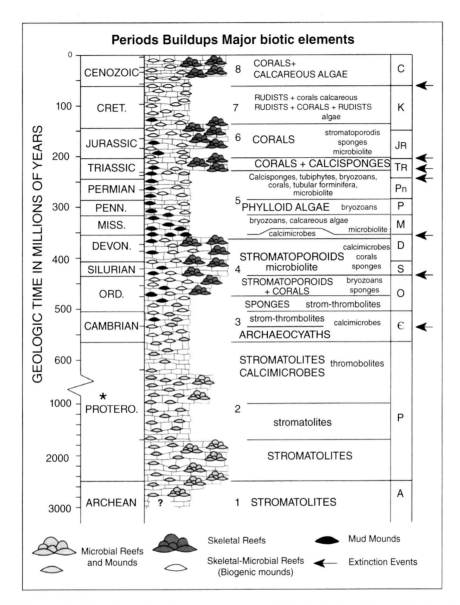

Figure 12.17 History of buildups through geological time. This is an idealized stratigraphic column representing geologic time and illustrating periods when there were only microbial reefs and mounds (indicated by green symbols below the Cambrian) from periods with skeletal reefs (indicated by blue symbols above the Cambrian) and skeletal-microbial reefs (biogenic mounds) (indicated by white symbols above the Proterozoic) as well as mud mounds. Numbers indicate different associations of reef- and mound-building biota. Arrows signal major extinction events, * = scale change. From James and Wood (2010). Reproduced with permission from Geological Association of Canada.

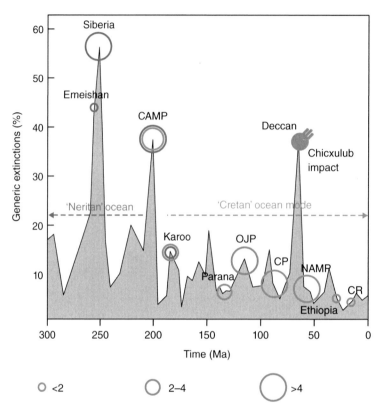

Figure 12.18 Mass extinctions and their relationship to large igneous provinces, the Neritan and Cretan oceans, and the Chicxulub impact. Mass extinctions are indicated by significant increases in generic extinction rates. Occurrences of large igneous provinces are indicated by circles overlying the extinction rate curve; the scale of circle sizes for large igneous provinces is in millions of cubic kilometers of basalt. Siberia, Siberian Traps; CAMP, Central Atlantic Magmatic Province; Deccan, Deccan traps; NAMP, North Atlantic Magmatic Province; OJP, Ontong Java; CP, Caribbean Plateaux; CR, Columbia River basalts. The end-Permian mass extinction occurred at the same time as the eruption of the Siberian traps, the end-Triassic mass extinction occurred at the same time as the eruption of CAMP, and the end-Cretaceous mass extinction is associated with the eruption of the Deccan traps and the Chicxulub impact. Note that eruption of large igneous provinces when there is a Cretan ocean did not lead to increases in extinction rate as compared to the Siberian trap and CAMP eruptions when a Neritan ocean existed, other than for the Deccan traps and the end-Cretaceous mass extinction, which may have been largely caused by the Chicxulub impact. From Sobolev et al. (2011). Reproduced with permission from Macmillan Publishers Ltd.

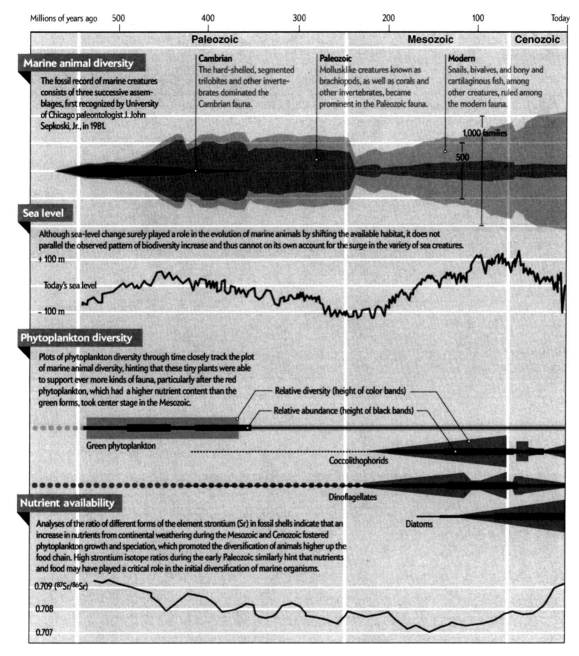

Figure 12.19 Mesozoic evolution of phytoplankton and the Modern Fauna. A mass extinction at the end of the Paleozoic, around 252 million years ago, preferentially affected the Paleozoic Fauna leading to the takeover by the Modern Fauna. The diversification of the Modern Fauna has been attributed to a variety of biotic factors as well as physical factors such as sea-level change. However, Mesozoic and Cenozoic increased nutrient flux into the oceans may have spurred the evolution of phytoplankton, the tiny plants that form the base of marine food chains, leading to increased diversification of the Modern Fauna. From Martin & Quigg (2013), where data sources are indicated.

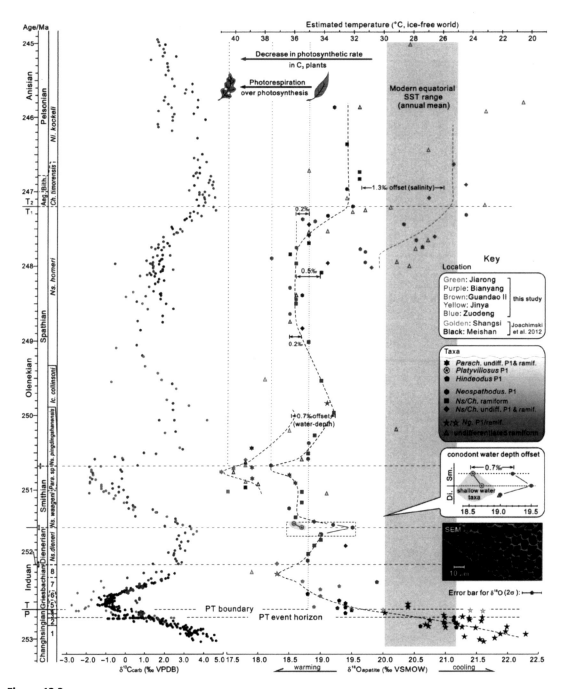

Figure 13.3

Figure 13.3 Sea surface temperature trends from the Late Permian into the Middle Triassic. These have been determined from oxygen isotopes of conodont apatite from the Nanpanjiang Basin of southern China and are compared with carbon isotopes of associated carbonates (see also Fig. 13.2). The estimated sea surface temperatures (SST) for this Permian-Triassic equatorial area are also compared with modern SST. Note near-synchronous perturbations in the records for both carbon and oxygen isotopes. Oxygen isotopes indicate a significant warming through the Permian-Triassic boundary transition, two thermal maxima in the late Griesbachian and late Smithian, cooling in the early Spathian followed by temperature stabilization, and further cooling and stabilization into the Middle Triassic. In the key location of data points of different colors is indicated; differently shaped icons show the different conodont taxa which were sampled for data points; and conodonts which lived at different water depths show offsets in estimated temperatures. Scanning electron microscope (SEM) analysis of conodont surfaces shows microreticulation and no sign of recrystallization, indicating a primary signal was obtained, as exemplified by the conodont surface SEM. At temperatures greater than 35 °C photorespiration predominates over photosynthesis, and above 40 °C most plants perish; leaf icons represent marine and terrestrial C3 plants. The occurrence of the end-Permian mass extinction (PT Event Horizon) occurs earlier than the biostratigraphically defined Permian-Triassic boundary (PT Boundary) (see Fig. 13.9). Aeg = Aegean; Bit. = Bithynian; for conodont zonations see original reference for abbreviations. Standards for isotope measurements: VSMOW = Vienna Standard Mean Ocean Water; VPDB = Vienna Pee Dee Belemnite. From Sun et al. (2012). Reproduced with permission from the American Association for the Advancement of Science.

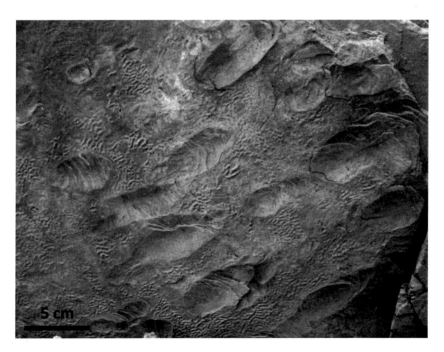

Figure 13.5 Lower Triassic (Spathian) wrinkle structures associated with unidirectional scour marks (scoop-shaped impressions) on a bedding plane of the Virgin Limestone Member (Moenkopi Formation). This bedding plane, found near Overton in the Muddy Mountains, Nevada (USA), forms the top of a calcareous siltstone. Wrinkle structures have crest heights of 1–2 mm, and associated sedimentary features include ripple marks and cross-bedding. Additional information can be found in Pruss et al. (2004). Photograph by Scott A. Mata. Reproduced with permission.

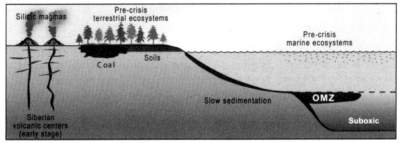

(a) Late Permian

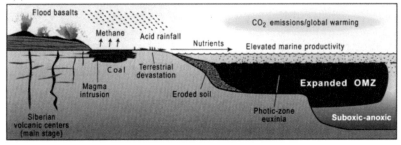

(b) End-Permian event

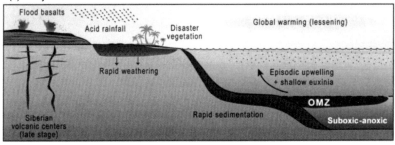

(c) Early Triassic

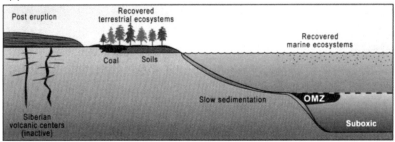

(d) Middle Triassic

Figure 13.11 Interpretive reconstructions of terrestrial–marine teleconnections from the Late Permian into the Middle Triassic. (a) Early stage Siberian Traps volcanism with minimal environmental effects during the Late Permian. (b) Main stage eruptions with attendant environmental effects during the latest Permian. (c) Late stage eruptions with lessening environmental effects during Early Triassic. (d) Post-eruption recovery of terrestrial and marine ecosystems in the Middle Triassic. From Algeo et al. (2011). Reproduced with permission from Elsevier.

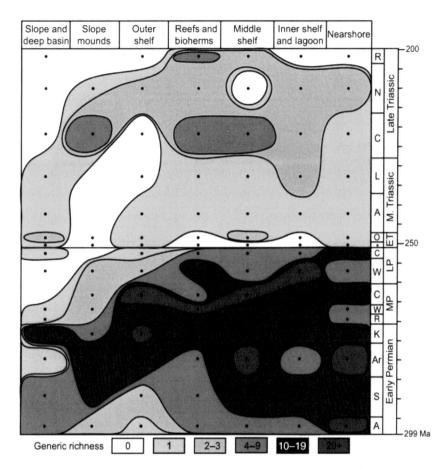

Figure 13.14 Contoured Permian–Triassic time–environment (T–E) diagram of marine stenolaemate bryozoans. Each dot represents a data point, either for an assemblage with the greatest bryozoan generic richness in each T–E bin, or for absence of bryozoans validated by the taphonomic control group (rhynchonelliform brachiopods). Note that bryozoan generic richness began to decrease in offshore environments in the Middle Permian, and this pattern of generic decrease continued into the Late Permian in more onshore environments. The end-Permian mass extinction then marked a geologically sudden extinction across all environments. Bryozoan generic richness was then very low until the Late Triassic when it began to increase in slope mound, reef, bioherm and middle shelf environments. From Bottjer et al. (2008). Reproduced with permission from the Geological Society of America.

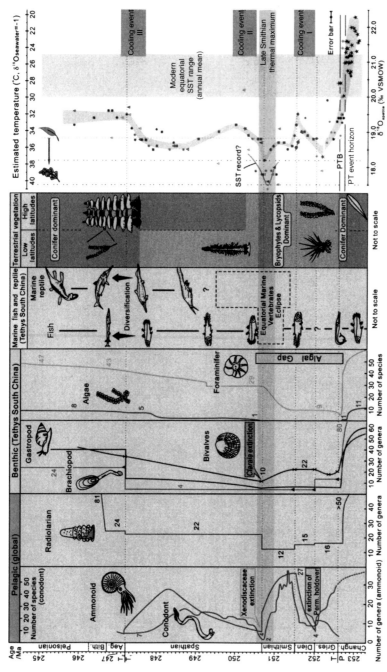

Figure 13.15 Early Triassic diversity of major marine groups and temperature trends. This shows an inverse relationship – peak diversity corresponds to cool climate conditions around the Dienerian – Smithian boundary, early Spathian, and early Anisian (named cooling events I to III), whereas low diversity in Griesbachian and Smithian correlates with peak temperatures. Fish and marine reptiles only show the general presence of taxa; no quantitative diversity data are available. Floral data show the loss of equatorial conifer-dominated forests above the Permian – Triassic (PT) boundary, with the earlier reappearance of this forest type at high latitudes. For estimated temperature (see also Fig. 13.3) vertically trending blue band represents the first-order seawater temperatures trend (upper water column, ~70 m water depth) whereas the red line indicated by "SST record?" represents possible sea surface temperatures derived from shallow water conodont taxa. Same stratigraphic scheme as in Fig. 13.3. From Sun et al. (2012). Reproduced with permission from the American Association for the Advancement of Science.

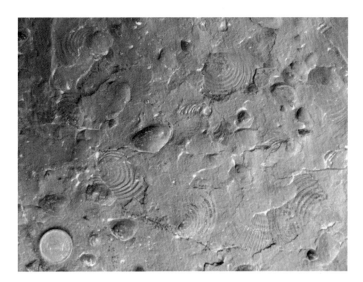

Figure 13.16 A characteristic Early Triassic (Induan) benthic assemblage consisting of the bivalves *Claraia* and *Unionites* on a bedding plane from the Lower Siusi Member of the Werfen Formation, northern Italy. *Claraia*, 2–4 cm in greatest dimension, has concentric growth lines and *Unionites* has a smoother more triangular shell. For life habits of *Claraia* and *Unionites* see Fig. 13.17. From the type location of the Siusi Member due south of the town of Siusi in the Dolomites of Alto Adige (South Tyrol). Photograph by Richard Twitchett. Reproduced with permission.

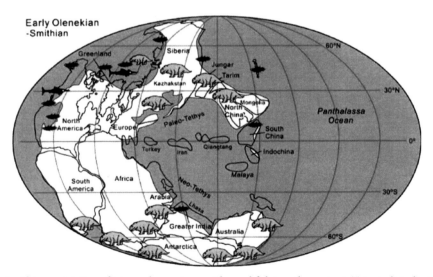

Figure 13.21 Smithian restriction of tetrapods, marine reptiles and fish to polar regions. Tetrapods indicated by gray symbols, fish and marine reptiles indicated by black symbols. From Sun et al. (2012). Reproduced with permission of the American Association for the Advancement of Science.

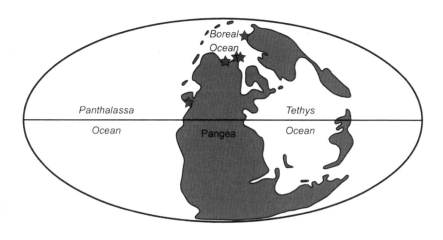

Figure 13.23
Paleobiogeographic distribution of Early Triassic bryozoans. These are indicated by stars, note concentration at north polar region. From Bottjer et al. (2008). Reproduced with permission of the Geological Society of America.

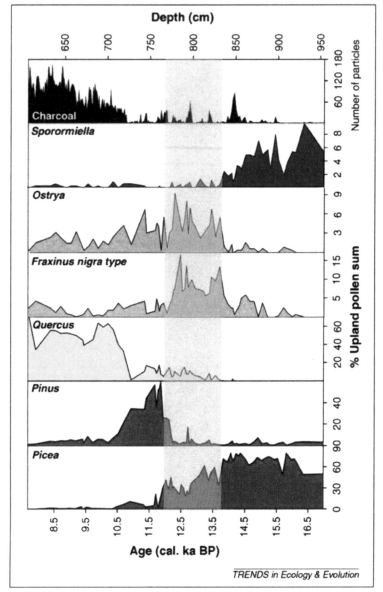

Figure 14.2 Development of non-analogue terrestrial communities. This is a summary pollen diagram from Appleman Lake, Indiana (US) for the period 8000–17000 years ago, when there was a global transition from glacial to interglacial conditions. Only the major tree pollen types are shown. Pollen assemblages with non-analogue modern pollen assemblages occur within 13,700–11,900 years ago (gray-shaded area). The percentage of spores of the dung fungus *Sporomiella*, a record of mega-faunal presence, and number of charcoal particles, a record of fire frequency and extent, are also shown. From Willis et al. (2010). Reproduced with permission from Elsevier.

Figure 14.5 Pangea and the areal distribution of CAMP volcanism. This also shows the location of major localities where CAMP and the Triassic–Jurassic mass extinction have been studied. (1) St. Audrie's Bay; (2) Newark basin; (3) Hartford basin; (4) Kennecott Point; (5) Val Adrara; (6) Moroccan CAMP sections. From Whiteside et al. (2010). Reproduced with permission from the National Academy of Sciences.

Figure 14.6 The end-Triassic extinction and extinction severity for organisms with an affinity for carbonate environments. (a) Phanerozoic biodiversity, note that the end-Triassic extinction represents perhaps the most severe Phanerozoic biodiversity crisis for the Modern Evolutionary Fauna. (b) Extinction rates through the Middle and Late Triassic to the Early Middle Jurassic. Extinction rates of reef genera and level-bottom (non-reef) genera that have an affinity for carbonate substrates are compared with the overall extinction rates of benthic invertebrate taxa. Note how Rhaetian (R) extinction rates are much higher for both carbonate level bottom faunas and reef faunas than the overall extinction rate and background levels. Extinction rates are binned by stage and therefore plotted at the midpoint of each age interval. From Greene et al. (2012). Reproduced with permission from Elsevier.

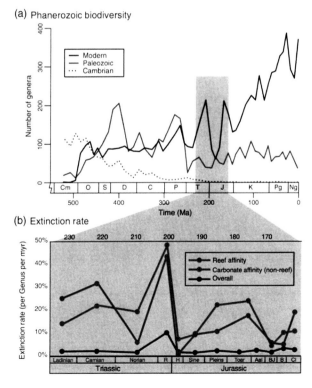

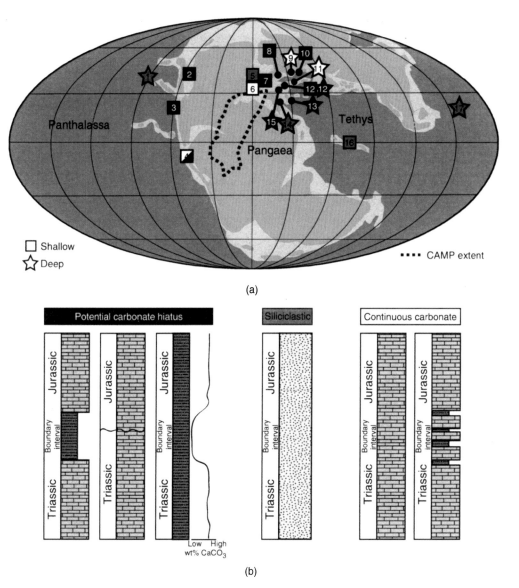

Figure 14.8 Reduction of carbonate deposition during the end-Triassic mass extinction. This is illustrated through examination of global lithological changes in Triassic–Jurassic boundary sections. (a) Early Jurassic paleogeographic reconstruction with the hypothesized extent of CAMP flood basalts (dashed outline) and the approximate paleolocation of Triassic–Jurassic boundary sites plotted. Squares represent shallow-water (shelf) sections while stars represent deep-water (basinal) sections. Dark brown indicates a potential carbonate hiatus across the boundary, light brown indicates predominantly siliciclastic sections, and white indicates either a section with potentially continuous carbonate deposition across the boundary or a section without reliable microstratigraphic data. (b) Idealized stratigraphic columns of each section type (carbonate hiatus, siliciclastic, and potentially continuous carbonate). 1, Queen Charlotte Islands, British Columbia, Canada; 2, Williston Lake, British Columbia, Canada; 3, New York Canyon, Nevada, USA; 4, Chilingote, Utcubamba Valley, Peru; 5, St. Audrie's Bay, England; 6, Asturias, northern Spain; 7, Mingolsheim core, Germany; 8, Northern Calcareous Alps, Austria; 9, Csovar section, northern Hungary; 10, Tatra Mountains (Carpathians), Slovakia; 11, Tolmin Basin (Southern Alps), Slovenia; 12, Lombardian Basin (Southern Alps), Italy; 13, Budva Basin (Dinarides), Montenegro; 14, Southern Apennines, Italy; 15, Northern and Central Apennines, Italy; 16, Germig, Tibet; 17, Southwest Japan. From Greene et al. (2012). Reproduced with permission from Elsevier.

1 Overview

Introduction

Paleoecology is the study of ancient ecology in its broadest sense. It has been enormously successful in placing the history of life within an ecological context. As part of that understanding, it has served as a vital tool for understanding the occurrence of many natural resources. In all its sophisticated approaches, paleoecology has taught us much about the past history of life and Earth's environments. With this record of demonstrating the response of Earth's biota to past environmental change, paleoecology now stands poised as a vital source of information on how Earth's ecosystems will respond to the current episode of global environmental change.

History of study

The notion that certain objects that one finds in sedimentary rocks were once living organisms is one that humanity struggled with for a long time. Leonardo da Vinci is generally credited with being the first to write down observations on the biological reality of fossils through examination of marine fossils from the Apennine Mountains of Italy. In reality, Leonardo also made some of the first paleoecological interpretations through

understanding these fossils as the remains of once living organisms that had not been transported some great distance and hence were not deposited as part of a great flood. The great utility of fossils to geologists was highlighted in the 19th century by the development of the geological timescale, and of course, after publication of "On the Origin of Species" by Darwin, evidence from the fossil record was some of the strongest available then for evolution. For the past 200 years, stratigraphic and paleontologic work has defined the occurrence of the major fossil groups that make up the record, and this general outline can be seen in Fig. 1.1, which shows Paleozoic, Mesozoic, and Cenozoic characteristic marine (ocean) skeletonized fossils.

Paleoecology as originally practiced is the use of biological information found in sedimentary rocks to help determine ancient paleoenvironments. Phanerozoic sedimentary rocks are found to have *in situ* marine fossils that we know were deposited in ancient oceans. Devonian and younger sedimentary strata that have remains of plants can be interpreted as deposited in terrestrial environments. For example, Fig. 1.2 shows the distribution within environments of various different fossil groups that have a substantial fossil record. One can see that these data are very valuable for understanding the past and past environments. So this information makes it easy to determine depositional environments

Paleoecology: Past, Present and Future, First Edition. David J. Bottjer.
© 2016 John Wiley & Sons, Ltd. Published 2016 by John Wiley & Sons, Ltd.

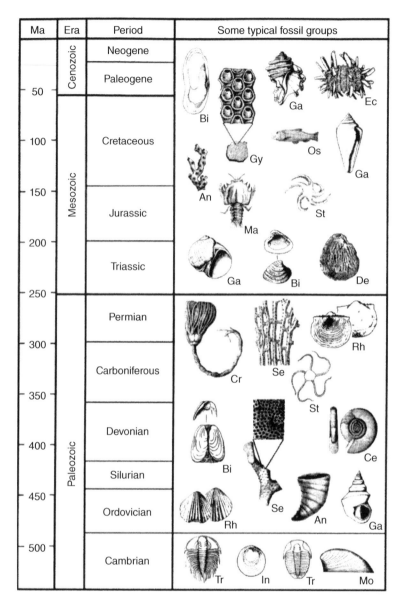

Figure 1.1 The Phanerozoic timescale with distribution of characteristic skeletonized marine fossils. Occurrence of fossils through the stratigraphic record has largely been determined through mapping efforts around the globe to characterize the surface geology of the continents. These fossil distributions have been continuously refined through the use of fossils to build the relative timescale and definition of Eras, Periods, and other time intervals. Key to classes: An, Anthozoa; Bi, Bivalvia; Ce, Cephalopoda; Cr, Crinoidea; De, Demospongiae; Ec, Echinoidea; Ga, Gastropoda; Gy, Gymnolaemata; In, "Inarticulata" (Linguliformea and Craniformea); Ma, Malacostraca; Mo, Monoplacophora; Os, Osteichthyes; Rh, "Articulata" (Rhynchonelliformea); Se, Stenolaemata; St, Stelleroidea; Tr, Trilobita. From McKinney (2007). Reproduced with permission from Columbia University Press.

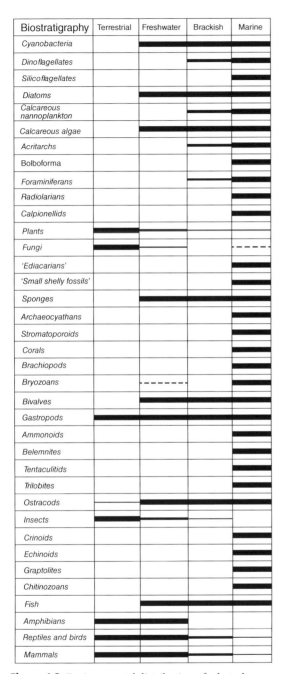

Figure 1.2 Environmental distribution of selected groups of fossils. This information largely comes from studies on the distribution of these organisms in modern environments, but also includes data on facies associations and functional morphology, particularly for the extinct groups. From Jones (2006). Reproduced with permission from Cambridge University Press.

of Phanerozoic sedimentary rocks, particularly in combination with physical sedimentary structures and geochemical indicators. Much work on paleoecology has been spurred by the petroleum industry and the need to understand ancient environments from drill cores and cuttings as well as outcrops. This need has led to much activity on microfossils, which can yield many specimens from a small piece of rock. And, through microfossils, information can be gained not only on ancient environments but also for ancient age determinations.

In the 1960s and 1970s, the study of fossil communities, or paleocommunities, blossomed. To many, the results from this research activity seemed to show that animals in the past lived the way they do today. But, as this information has accumulated, it became clear that ecology changes through time, due to both evolution as well as environmental change. The synthesis of this realization has come to be known as evolutionary paleoecology. Evolutionary paleoecology has become a group of research programs that focus on the environmental and ecological context for long-term macroevolutionary change as seen from the fossil record. For example, Fig. 1.3 displays the tiering history for benthic suspension-feeding organisms in shallow marine environments below wave base since their early evolution in the Ediacaran, synthesized in work done with William Ausich. Tiering is the distribution of organisms above and below the seafloor, and this diagram shows how the distribution has changed through time and therefore how organisms have evolved their ability to inhabit three-dimensional space. This diagram is the latest of several showing tiering, and its development in the early 1980s was part of the early history of evolutionary paleoecology.

Paleoecology and the future

Earth's ancient ecology is a fascinating subject for study, but there is more to be gained from this study as a benefit to present society. We are entering a time of widespread environmental change, in large

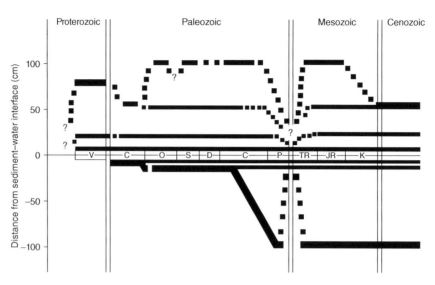

Figure 1.3 Tiering history among marine soft-substrata suspension-feeding communities from the late Precambrian through the Phanerozoic. Zero on the vertical axis indicates the sediment–water interface; the heaviest lines indicate maximum levels of epifaunal or infaunal tiering; other lines are tier subdivisions. Solid lines represent data, and dotted lines are inferred levels. These characteristic tiering levels were determined for infaunal tiers by examination of the trace fossil record, particularly the characteristic depth of penetration below the seafloor of individual trace fossils. Data on shallow infaunal tiers also came from functional morphology studies of skeletonized body fossils. Paleocommunity and functional morphology studies of epifaunal body fossils comprise the data for epifaunal tiering trends. Tiering data from the late Precambrian is from studies of the Ediacara biota. This tiering history has been updated as more data have become available. From Ausich and Bottjer (2001). Reproduced with permission from John Wiley & Sons.

part due to disruption of the carbon cycle (Fig. 1.4) through burning of lithospheric coal and petroleum and subsequent transfer of carbon in the form of carbon dioxide from the lithosphere into the atmosphere. This increase in greenhouse gasses in the atmosphere is causing rapid increased warming of the atmosphere and the ocean (Fig. 1.5). Increased warming of the ocean can lead to reduced ocean circulation which causes decreased oxygen content in ocean water and hence the growth of ocean systems characterized by reduced to no oxygen content, called "dead zones" (Fig. 1.6). Increased levels of atmospheric carbon dioxide cause decreases in the concentration of the carbonate ion in ocean water, termed ocean acidification, which makes it more difficult for many organisms such as corals to produce their calcium carbonate skeletons (Fig. 1.7).

As is discussed in later chapters, the fossil record contains evidence for a wide variety of past environmental changes, some of which are strikingly similar to current anthropogenically created changes. Thus, Earth has run the experiment in the past of what happens when there is an episode of geologically sudden global warming, termed a hyperthermal. The ecological changes that occurred during these ancient episodes can be studied to help provide data which can help manage our future interval of environmental change. This approach has been broadly developed under the new field of conservation paleobiology. In particular, one major aspect of conservation paleobiology is conservation paleoecology, which focuses on providing data from the past to manage future ecological changes.

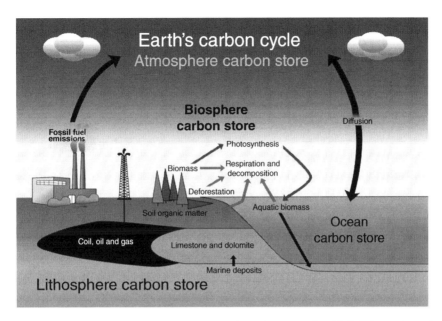

Figure 1.4 Schematic of modern carbon cycle including anthropogenic influence. Combustion of lithospheric carbon such as coal and oil is the modern cause of global warming, and a similar mechanism involving igneous intrusions through sedimentary rocks rich in carbon has been the cause of rapid global warming episodes, or hyperthermals, in the past. From the New York State Department of Environmental Conservation website: http://www.dec.ny.gov/energy/76572.html. (*See insert for color representation.*)

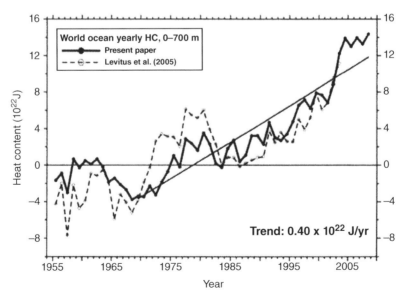

Figure 1.5 Increase in ocean heat content since 1955 shown as a time series of yearly ocean heat content in joules (J) for the 0–700 m layer. Each yearly estimate is plotted at the midpoint of the year, with the reference period from 1957 to 1990. From Levitus et al. (2009). Reproduced with permission from John Wiley & Sons.

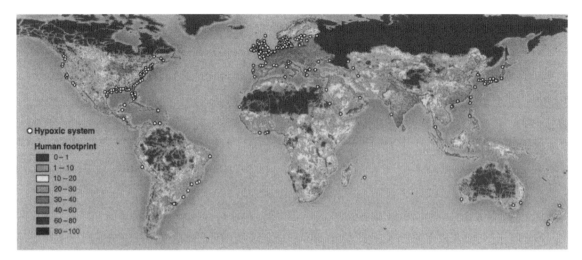

Figure 1.6 Location of hypoxic system coastal "dead zones." Their distribution matches the global human footprint, where the normalized human influence is expressed as a percent, in the Northern Hemisphere. For the Southern Hemisphere, the occurrence of dead zones is only recently being reported. From Diaz and Rosenberg (2008). Reproduced with permission from the American Association for the Advancement of Science. (*See insert for color representation.*)

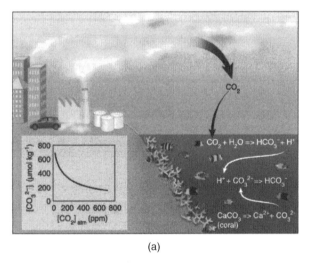

(a)

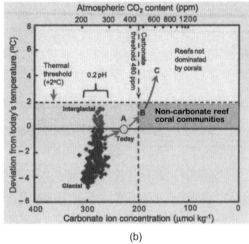

(b)

Figure 1.7 Increase in atmospheric carbon dioxide and its influence on ocean acidification and the resultant affect on development of coral reefs in the past, present, and future. (a) Increased carbon dioxide concentration in the oceans leads to decreased availability of carbonate ions, which are needed by corals to secrete their skeletons made of calcium carbonate. (b) Plot of temperature, atmospheric carbon dioxide content, and ocean carbonate ion concentration showing the predicted trend in the future of reefs not dominated by corals with increased levels of acidification. From Hoegh-Gulberg et al. (2007). Reproduced with permission from the American Association for the Advancement of Science. (*See insert for color representation.*)

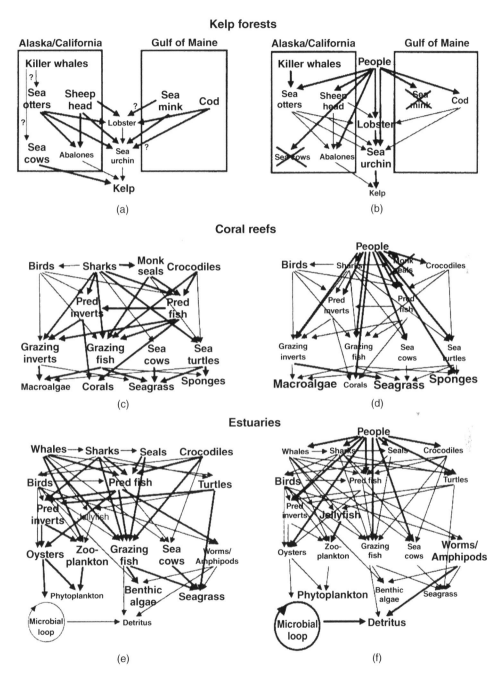

Figure 1.8 The effects of human overfishing on coastal ecosystems. Simplified food webs showing changes in some of the important top–down trophic interactions before and after fishing in kelp forests, coral reefs, and estuaries. Bold font represents abundant, normal font represents rare, "crossed out" represents extinct, thick arrows represent strong interactions, and thin arrows represent weak interactions. From Jackson et al. (2001). Reproduced with permission from the American Association for the Advancement of Science.

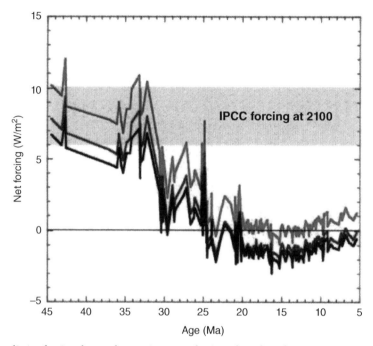

Figure 1.9 The net radiative forcing due to changes in atmospheric carbon dioxide concentration and total solar irradiance from 5 to 45 million years ago. The three curves represent the range in carbon dioxide concentration using three different proxies for ancient atmospheric carbon dioxide. The shaded area denotes the range in radiative forcing projected to occur by 2100 according to the Intergovernmental Panel on Climate Change (IPCC) Fourth Assessment Report (2007). Net radiative forcing is in watts per square meter. From Kiehl (2011). Reproduced with permission from the American Association for the Advancement of Science. (*See insert for color representation.*)

Along with the environmental changes that are created by global warming, we also see other anthropogenic effects such as increased runoff of nutrients from human activity, which has spurred the growth of dead zones in coastal ecosystems (Fig. 1.6). Along with increased hypoxia, modern ocean ecosystems are also impacted by the anthropogenic effects of overfishing. Figure 1.8 shows the change in trophic webs that has occurred from times before intensive human fishing to after fishing in environments such as kelp forests, coral reefs, and estuaries. These sorts of impacts can also be studied and managed for the future by studying paleoecology of the last few thousand years to understand how human impact has changed these ecosystems and present another aspect of conservation paleoecology.

The import of studying past environmental change and its impact on ecosystems can be viewed through a recent study done by Jeff Kiehl (Fig. 1.9). This study calculates the net forcing in watts per square meter from 5 to 45 million years ago, using three different proxies for carbon dioxide concentration in the atmosphere. Forcing decreased from a greenhouse climate 35–45 million years ago to an icehouse climate like the one today 20–25 million years ago, with extensive ice at the poles. Also plotted is the range of net forcing that the Intergovernmental Panel on Climate Change (IPCC) report of 2007 forecasted for the year 2100. This range is the same as 35–45 million years ago, which implies that in 100 years, the Earth's ecosystems will journey from an icehouse to a greenhouse climate. The rapidity of this change is dramatic when compared with the 10–15 million years that elapsed during the Cenozoic transition from greenhouse to icehouse. It remains to be seen how Earth's ecosystems will respond to this projected episode of hyperthermal climate change, and conservation paleoecology may provide a key to managing the future.

Summary

Paleoecology has deep roots that were initiated with humankind's understanding that fossils are natural objects that provide evidence on ancient ecosystems. This is a vast subject that has only minimally been addressed, as Earth's environments have changed dramatically though the long history of life on this planet, and evolutionary changes in response to these environmental changes have been complex and varied. Within this storehouse of evidence on ecosystem response to environmental change that is available in the fossil and stratigraphic record lie many clues on how we can manage the current episode of global ecosystem change.

References

Ausich, W.I. & Bottjer, D.J. 2001. Sessile Invertebrates. *In* Briggs, D.E.G. & Crowther, P.R. (eds.), *Palaeobiology II*. Blackwell Science, Oxford, UK, pp. 384–386.

Diaz, R.J. & Rosenberg, R. 2008. Spreading dead zones and consequences for marine ecosystems. *Science* 321, 926–929.

Hoegh-Gulberg, O., Mumby, P.J., Hooten, A.J., Steneck, R.S., Greenfield, P., Gomez, E., Harvell, C.D., Sale, P.F., Edwards, A.J., Caldeira, K., Knowlton, N., Eakin, C.M., Iglesias-Prieto, R., Huthiga, N., Bradbury, R.H., Dubi, A. & Hatzioios, M.E. 2007. Coral reefs under rapid climate change and ocean acidification. *Science* 318, 1737–1742.

Jackson, J.B.C., Kirby, M.X., Berger, W.H., Bjkorndal, K.A., Botsford, L.W., Bourque, B.J., Bradbury, R.H., Cooke, R., Erlandson, J.K., Estes, J.A., Hughes, T.P., Kidwell, S., Lange, C.B., Lenihan, H.S., Pandolfi, J.M., Peterson, C.H., Steneck, R.S., Tegner, M.J. & Warner, R.R. 2001. Historical overfishing and the recent collapse of coastal ecosystems. *Science* 293, 629–637.

Jones, R.W. 2006. *Applied Palaeontology*. Cambridge University Press, Cambridge, UK.

Kiehl, J. 2011. Lessons from Earth's past. *Science* 331, 158–159.

Levitus, S., Antonov, J.I., Boyer, T.P., Locarnini, R.A., Garcia, H.E. & Mishonov, A.V. 2009. Global ocean heat content 1955-2008 in light of recently revealed instrumentation problems. *Geophysical Research Letters* 36, doi:10.1029/2008GL037155.

McKinney, F.K. 2007. *The Northern Adriatic Ecosystem: Deep Time in a Shallow Sea*. Columbia University Press, New York.

Additional reading

Dietl, R.G. & Flessa, K.W. (eds.). 2009. Conservation Paleobiology: Using the Past to Manage for the Future. *The Paleontological Papers*, Volume 15, The Paleontological Society.

Solomon, S., et al. (eds.). 2007. *Climate Change 2007: The Physical Science Basis. Contribution of Working Group I to the Fourth Assessment Report of the Intergovernmental Panel on Climate Change*. Cambridge University Press, Cambridge, UK.

2 Deep time and actualism in paleoecological reconstruction

Introduction

The perception and appreciation of time is a difficult topic for human beings. We are aware of long intervals of time on the human scale because we are taught human history. And we have short-term and long-term strategies in making plans, although typically when there is a choice, we pick the short-term solution to a problem. Our time perception as a species is strongly molded by our evolutionary context, particularly our generation time. Only with the rise of science have we accumulated data empirically that have allowed us to understand that there not only is historical time but that there are billions of years that have transpired in Earth's deep time history.

Perceptions of time

This dichotomy I find in my own personal experience. When I was a grade school student, I learned about history back to ancient Egypt and Mesopotamia. In my mind, the beginning of these ancient civilizations in the Middle East on my historical timescale all seemed like very long ago. But then in college, I learned about the geological timescale and how deep in time it goes, and my appreciation of the historical timescale changed to an understanding that it represents just that last tiny bit of the geological timescale.

This is the conundrum that we wrestle with when interacting with the public. When we speak of time, most people really have that historical timescale on their minds, while we're talking about geological timescales. So, when we say there was a "Cambrian explosion," it is hard for many people to understand that we are talking about geological time and that an evolutionary process that happened over several million years of geological time can be described as an explosion.

Our ability to measure deep time precisely has been improving by leaps and bounds since the discovery of radioactivity and the development of radiometric age dating. For the deep time record of paleoecology to be relevant for managing future environmental change, we need to separate processes that are observed in deep time into ones that act on scales such as our current rate of change of decades to hundreds of years from those that act over millions of years. The relevance to modern society of a large asteroid hitting the Earth was discovered through study of the rock and fossil record because the effects of an impact occur over human timescales. Processes that occur over millions of years are intrinsically fascinating but not likely to be relevant to understanding and managing the environmental problems that society currently faces.

Paleoecology: Past, Present and Future, First Edition. David J. Bottjer.
© 2016 John Wiley & Sons, Ltd. Published 2016 by John Wiley & Sons, Ltd.

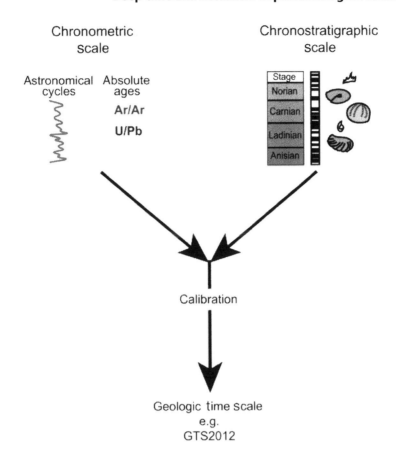

Figure 2.1 Steps in construction of a geological timescale. The chronostratigraphic scale is a relative timescale and includes (from left to right) formalized definitions of geologic stages (here with examples of Triassic stages), magnetic polarity zones, and biostratigraphic zonation units, with examples here indicated by fossil symbols (from top to bottom, conodont, ammonoid, echinoderm, foraminifera, bivalve). The chronometric scale is measured in years and includes absolute ages measured from radiogenic isotope systems such as argon/argon and uranium/lead and astronomical cycles exemplified by the sedimentary expression of Earth's orbital cycles. These orbital cycles, termed Milankovitch cycles, have specific time implications and can be detected from measurements on sedimentary bed thickness, composition, and geochemistry. Other chronostratigraphic approaches not illustrated here include stable isotope stratigraphy (strontium, osmium, sulfur, oxygen, carbon). Commonly, when determining the age of a sedimentary section, fossils can be collected for biostratigraphic determinations. Fossils, along with sedimentary samples, can be analyzed for geochemical data, and other sedimentary observations can be made for determination of astronomical cycles (if this evidence is available). Oriented sedimentary samples can also be collected for analysis in a magnetometer to detect reversed and normal polarity zones. If volcanic rocks, such as tuffs, exist, these can be sampled for radiogenic isotope measurements to determine an absolute age in years. If some or all of this chronometric and chronostratigraphic information is available, it can then be merged to produce an age calibration that allows linkage into a formal geologic timescale, here indicated as that found in Gradstein et al. (2012). From Gradstein et al. (2012). Reproduced with permission from Elsevier. (*See insert for color representation.*)

Geological time

The great age of the Earth was appreciated long before any numerical dates could be reliably calculated. Using Steno's laws of superposition, original horizontality, and lateral continuity, the relative timescale was developed, extended, and filled out in detail. Names rather than dates for the various levels of this relative chronostratigraphic timescale were a necessity, because no one could determine how old they really were. Even today, most geologists prefer to use relative time terms rather than numerical ages when communicating with each other.

As shown in Fig. 2.1, to get the numerical age of most geologic events, the geologist must first determine its relative age by fossils using the chronostratigraphic scale. Then this scale is compared with the chronometric scale of absolute ages from which the calibration is made. From thousands of these kinds of determinations, the geological timescale has been constructed. As shown in Fig. 2.2, additional approaches have also been useful for developing the geological timescale at different intervals throughout the Phanerozoic. Such techniques as magnetostratigraphy as well as orbital forcing analyses are included in a variety of other approaches that have been valuable for certain time intervals.

Everyone in the earth sciences remembers memorizing the geological timescale with the accompanying absolute ages from their introductory geology class in college. And most of us remember there is much underpinning the timescale, starting with the relative age of sedimentary units first understood in the 19th century. It may have seemed that the absolute ages assigned to the timescale are static, but in fact, there is much ongoing work providing greater and greater precision to the timescale in an attempt to drive toward provision of information from deep time on rates of processes that occur over periods of time as short as a human timescale. This effort is exemplified by the large integrated project EARTHTIME, which is constantly pushing the technological envelope to provide increased precision of absolute ages. The

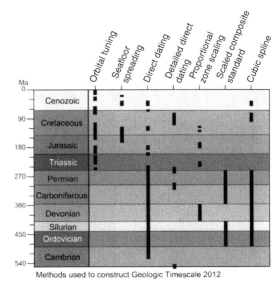

Figure 2.2 Methods used to construct the geological timescale for the Phanerozoic in Gradstein et al. (2012), which depend on the quality of data available for different time intervals. Cyclostratigraphic analyses of Milankovitch orbital cycles are used in orbital tuning approaches. Seafloor spreading rates are calculated from the distribution of ocean seafloor magnetic anomalies. Direct dating involves use of high-precision radiogenic isotope ages, usually determined from zircons collected from volcanic rock. Proportional zone scaling and scaled composite standard analyses involve scaling using biostratigraphic data. Cubic spline curve fitting geomathematically relates observed ages to their stratigraphic position. From Gradstein et al. (2012). Reproduced with permission from Elsevier. (*See insert for color representation.*)

current geological timescale shown in Fig. 2.3 is stable in some intervals and very stable in other intervals but is still undergoing extensive work to refine ages and stratigraphic relationships.

Uniformitarianism and actualism

In the 18th century, the Scottish geologist James Hutton recognized that rocks exposed on the Earth's surface were the product of continuing Earth processes, rather than a single supernatural creation

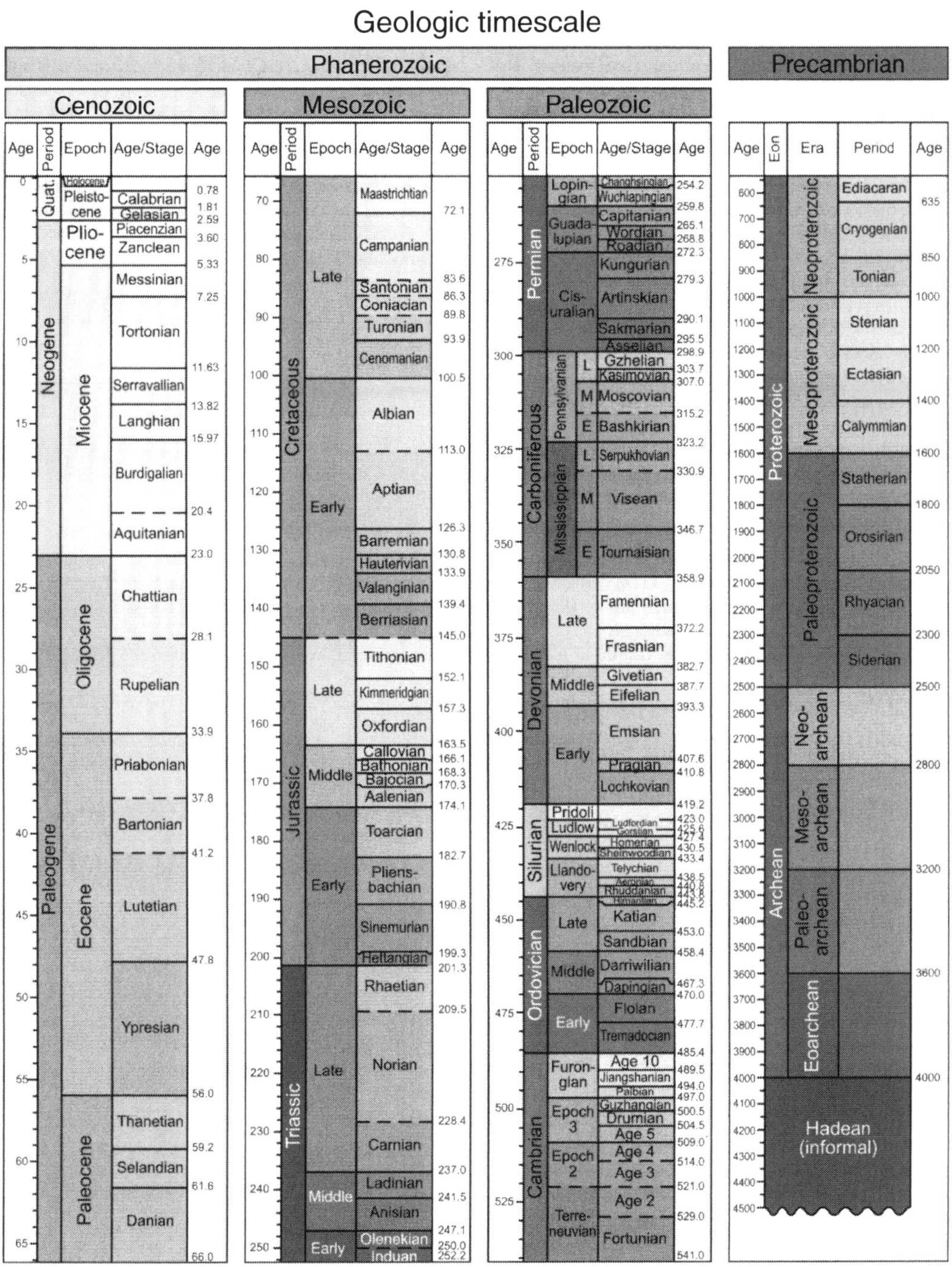

Figure 2.3 Geological timescale From Gradstein et al. (2012). Reproduced with permission from Elsevier. (*See insert for color representation.*)

or Noachian deluge. This concept is called naturalism or uniformitarianism, and it is a methodology of inferring ancient events and environments by analogy with processes observable in the modern world. In contrast, the viewpoint that prevailed before Hutton is termed catastrophism, because its adherents proposed supernatural explanations such as a catastrophic global flood to explain evidence found in the rock record.

In the 19th century, Hutton's approach to reconstructing Earth's history achieved wide acceptance through the work of Charles Lyell. To combat the catastrophists, Lyell took an extreme position on uniformitarianism and so rejected all interpretations that included catastrophic processes in Earth's history. Lyell's gradualist bias was so strong that for generations it influenced geologists who were reconstructing Earth's history to deny strong evidence for natural catastrophes different or larger than those known from observations on the modern Earth. Certainly, large asteroid or comet impacts are not easily accommodated by a gradualistic scenario of slow, steady, cumulative change. Yet, as we have discovered in the last 30 years, it appears that many rapid and sometimes unique events have had a major effect on the fossil and stratigraphic record. Nevertheless, actualism is a methodological assumption that is critical to all of the historical natural sciences.

The two terms uniformitarianism and actualism are both commonly used interchangeably worldwide, although actualism is used more regularly in continental Europe. Although the usual approach for reconstructing history in the natural world uses actualism as a dominant guiding principle, reconstruction of Earth's biological history requires a different approach from the use of immutable physical and chemical axioms. The reason for this difference is because biological features of Earth's

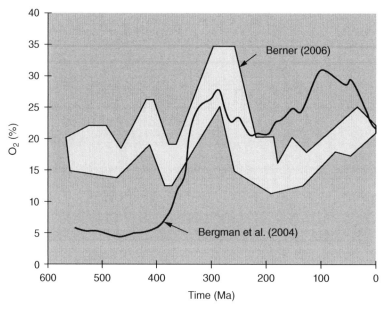

Figure 2.4 Estimates of Phanerozoic atmospheric O_2 concentrations from two different models, showing Paleozoic O_2 peak in the Carboniferous. These O_2 curves are produced using biogeochemical models, the Carbon-Oxygen-Phosphorus-Sulfur-Evolution (COPSE) model by Bergman et al. (2004) and the GEOCARBSULF model by Berner (2006). Model inputs include carbon and sulfur weathering and burial rates, and different model assumptions lead to different oxygen concentrations and the different oxygen curves shown here for the Mesozoic and Cenozoic. From Kasting and Canfield (2012). Reproduced with permission from John Wiley & Sons.

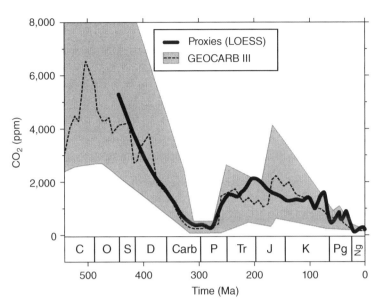

Figure 2.5 Atmospheric CO_2 through the Phanerozoic, reconstructed using proxies for CO_2 and GEOCARB III, a biogeochemical carbon cycle model developed by Berner and Kothavala (2001). Proxies for CO_2 include stomatal densities and indices in plants, the δ C13 of soil minerals, and the δ B11 of marine carbonates. Smoothed proxy data is plotted using a locally weighted regression (LOESS). The best-guess predictions of GEOCARB III are plotted as a dashed line, and the range of reasonable predictions of this model are shown as a gray-shaded region. From Royer (2006). Reproduced with permission from Elsevier.

Figure 2.6 Reconstruction of a Carboniferous forest including a dragonfly with a wingspan of 60 cm. Correlation of large insect size with atmospheric oxygen content assumes insect size limitation is related to the surface area of the respiratory system versus organism size, so that all other things being equal an increase in atmospheric oxygen content allows a larger body size. Other size limitations such as the lack of predators like birds or pterosaurs which had not yet evolved during this time have also been suggested as contributing to the large size of Carboniferous dragonflies. From Kump et al. (2009). (*See insert for color representation.*)

environments, by their very nature, have changed through time due to organic evolution.

For example, by the Devonian lignin had evolved to provide an important structural element that allowed trees to gain significant height. In the Carboniferous coal forests proliferated, burying a lot of carbon into the lithosphere as coal was formed. With this burial of carbon as coal and the withdrawal of carbon from the atmosphere, the percentage of oxygen in the atmosphere increased while carbon dioxide decreased, as has been modeled by several authors and is shown in Figs. 2.4 and 2.5. It is thought that this increase in atmospheric oxygen led to a variety of biotic effects, including the evolution of particularly large dragonflies the size of seagulls that are typical of Carboniferous forests, as shown in Fig. 2.6. This rise in atmospheric oxygen concentrations along with the drop in atmospheric CO_2 is dramatic.

Summary

The ability to peer into deep time to understand Earth's history has been one of humankind's most astounding accomplishments. Integrated over geological time, the effects of evolution on the history of Earth are dramatic. Thus, we have learned that ancient biological attributes of the environment no longer exist or are dominant in modern settings. In order to interpret such features in the ecological realm, one must adopt the view that an actualistic methodology will not solve all problems in paleoecology and that a nonactualistic approach sometimes provides keys to understanding ancient ecologies. Similarly, this viewpoint is also gaining acceptance in sedimentology, and appreciation of biogenic effects upon all processes, particularly during the early evolution of animals and plants, has begun to be studied in the context of nonactualism.

References

Bergman, N.M., Lenton, T.M. & Watson, A.J. 2004. COPSE: A new model of biogeochemical cycling over Phanerozoic time. *American Journal of Science* 304, 397–437.

Berner, R.A. 2006. GEOCARBSULF: A combined model for Phanerozooic atmospheric O_2 and CO_2. *Geochimica et Cosmochimica Acta* 70, 5653–5664.

Gradstein, F.M., Ogg, J.G., Schmitz, M.D. & Ogg, G.M. (eds.). 2012. *The Geologic Time Scale 2012*, Volumes 1 and 2. Elsevier, Amsterdam.

Kasting, J.F. & Canfield, D.E. 2012. The Global Oxygen Cycle. *In* Knoll, A.H., Canfield, D.E. & Kornhauser, K.O. (eds.), *Fundamentals of Geobiology, 1st Edition*. Wiley-Blackwell, pp. 93–104.

Kump, L.R., Kasting, J.F. & Crane, R.G. 2009. *The Earth System*. Prentice Hall, San Francisco.

Royer, D.L. 2006. CO_2-forced climate thresholds during the Phanerozoic. *Geochimica et Cosmochimica Acta* 70, 5665–5675.

Additional reading

Bottjer, D.J. 1998. Phanerozoic non-actualistic paleoecology. *Geobios* 30, 885–893.

3

Ecology, paleoecology, and evolutionary paleoecology

Introduction

Ecology is the study of the interactions between organisms and the Earth as well as between organisms. Thus there is autecology, which is concerned with individual organisms and how they function, and synecology, which considers interactions with other organisms and the surrounding physical and chemical environment. In the broad variety of environments on Earth there are groups of organisms that are adapted to particular physical and chemical conditions, and at the smallest level these are called communities. Communities group together on a larger scale at various biogeographic levels. How ecology has played out through time, with the influence of evolution and changing environments, is the subject of paleoecology.

Ecology and paleoecology

A lot of the work that has been done through the past centuries has been to determine the characteristics of different marine and terrestrial environments and how they can be classified (Fig. 3.1). For example, marine environments are categorized according to various water depths both in benthic or seafloor environments as well as in pelagic environments

in the water column (Fig. 3.1). Much ecological work such as this has led to the understanding of modern ecosystems and the feeding relationships, or trophic relationships, that are found there. An example of ocean ecosystem feeding relationships is shown in the schematic in Fig. 3.2. Energy flow passes through these communities with various producers utilizing photosynthesis as the prime source of captured energy for the consumer groups in nature. Some organisms in communities have larger roles through producing physical structure in the environment, such as reef builders or trees. These ecological engineers are termed keystone species. Keystone species can also be the most abundant in a community, where these dominants can control the energy flow through the system.

Paleoecologists are able to study various aspects of ancient ecology, depending upon the nature of the fossil and stratigraphic record that is being studied. The sediments in which fossils are contained include significant physical and chemical evidence on ecological interactions, and the fossils themselves provide evidence on the variety of life habits present. Clearly not all the evidence that one can gather on modern ecology is available through study of the fossil record. So, paleoecology is not a one-to-one match to ecology, where all one finds are imperfectly preserved examples of ecosystems better studied by ecologists. Because paleoecology has the

Paleoecology: Past, Present and Future, First Edition. David J. Bottjer.
© 2016 John Wiley & Sons, Ltd. Published 2016 by John Wiley & Sons, Ltd.

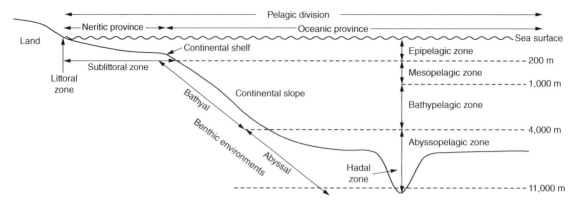

Figure 3.1 Categorization of Earth's environments. Marine environments, as shown in this schematic, are defined in a variety of ways with an emphasis on water depth. These include benthic (or seafloor) environments and pelagic environments in the water column. The littoral zone is the mosaic of shoreline environments. The continental shelf divides the benthic sublittoral and pelagic neritic from oceanic benthic (bathyal, abyssal, hadal) and pelagic (epipelagic, mesopelagic, bathypelagic, abyssopelagic) environments. Terrestrial environments differ according to variations in temperature, humidity, and elevation, and include freshwater environments such as wetlands, ponds, lakes, and streams, and subaerial environments such as deserts, grasslands, shrub lands, and forests. From Brenchley and Harper (1998).

advantage of allowing us to look deeply into time, as the themes in this text will portray, there are strong avenues of ecological evidence which run through the fossil record that offer ample opportunity to study unique aspects of the ways Earth's ecosystems have and can function.

Functional morphology

An important component of ecology as well as paleoecology is understanding how animals live and how their different morphological features operate, which constitutes the field of functional morphology. Interpreting how ancient organisms lived is one of the major activities of paleontology. There are three approaches to interpreting the life habits of ancient organisms from fossils: (1) comparison with modern analogs; (2) production of theoretical, computer, and physical models; and (3) circumstantial evidence.

When there are living descendants of the fossil organisms that are phylogenetically linked to the fossil organisms, interpretation of the function of the fossil is easily done. A good example of this is the study of the life habits of the bivalve mollusc group, the scallops. These intriguing animals live on the seafloor, with an ovate to teardrop-shaped shell that has extensions near the hinge line, termed the auricles. Many modern scallops are attached to an object on the seafloor by an organic tether called the byssus. What is hard to imagine is that a variety of modern scallops are swimmers, which they accomplish by clapping their valves and jetting water from the hinge side of the animal. This ungainly form of swimming appears to largely function as a means to escape from slow moving predators such as starfish. It would not be apparent that some scallops swim if all we had were fossil scallops. But, in a pioneering study (Fig. 3.3), Steve Stanley made observations of modern scallops which show that the swimming scallops have a broad shell with similarly sized auricles, for swimming efficiency. Stanley also found that attached scallops have asymmetrical auricles, as the larger auricle is used like an outrigger so that the byssus can hold the organism firmly to the substrate.

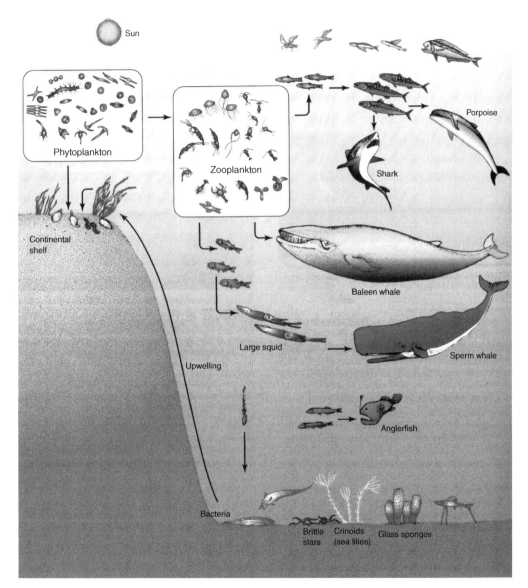

Figure 3.2 Marine food web schematic. Producers which capture energy from sunlight through photosynthesis include phytoplankton and seafloor plants on the continental shelf. Zooplankton consume the phytoplankton. Baleen whales and small fish consume the zooplankton, and the small fish are in turn consumed by squid and larger fish. Squid are consumed by sperm whales and the larger fish are consumed by sharks and porpoises. In the deep sea well below the photic zone the abundance of life is diminished with only a few seafloor consumers such as sponges and crinoids filtering plankton suspended in the seawater. A more common strategy for deep seafloor animals is to extract their food from sediment which they ingest while burrowing, known as deposit-feeding. In the deep sea larger fish such as anglerfish attract and consume smaller fish. Remains of organisms from above settle to the seafloor, to be decomposed by bacteria, and the nutrients left from this process are returned to the surface by upwelling for use again by photosynthetic life. From Stanley (2008). Macmillan Higher Education. (*See insert for color representation.*)

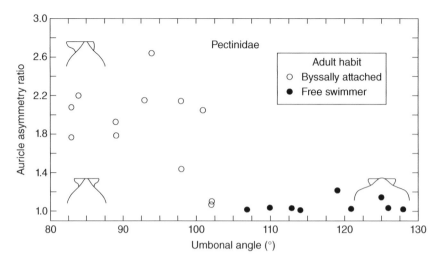

Figure 3.3 Morphologic features of byssally attached and free-swimming scallops. The characteristic shape of a scallop is a relatively flat pair of teardrop to ovate shells which articulate along a hinge line that is accentuated by a more pointed structure called the umbo. At each side of the umbo are triangularly shaped projections of the shell called the auricles. The pointed intersection of the shell that forms the umbo can be measured as the umbonal angle; a smaller umbonal angle produces a more teardrop-shaped shell, and a larger umbonal angle produces a more ovate shell. In a scallop species the size of the auricles can be equal or unequal. The degree to which they are unequal can be expressed by measuring the dimensions of the two auricles along the straight edge to the intersection with the umbo, and calculating their ratio. As shown, scallops with a relatively low umbonal angle and teardrop shape also have asymmetrical auricles, with the larger auricle acting as an outrigger for these byssally attached organisms, where the byssus is located at the intersection of the auricle with the main part of the shell away from the point of the umbo. Scallops with equal auricles typically have large umbonal angles and more ovate shapes which allows for better hydrodynamic behavior of the shells during swimming. From Stanley (1970). Reproduced with permission from the Geological Society of America.

These observations from modern scallops can then be used to interpret the functional morphology of fossil scallops.

Extinct organisms that have no direct evolutionary descendants pose additional problems, because there is no living analog. In that case models can be made of the ancient organism, which exhibit the function which that organism is thought to have displayed. These can be theoretical models, computer models, or actual physical models. Perhaps one of the best known examples of a physical model is the construction of models of the extinct pterosaurs, an example of which is shown in Fig. 3.4. This model was built for the Smithsonian Institution and was flown in Death Valley and other locales. Through making actual flying models of pterosaurs much has been learned about the dynamics of pterosaur flight.

Aspects of how ancient organisms lived are sometimes "frozen" in the fossil record. In particular this circumstantial evidence provides important clues on how extinct organisms lived. As discussed in Chapter 5, trace fossils are prime examples of information on behavior of fossil organisms. Well-preserved body fossils can also preserve unique biological information. For example, we know that Mesozoic marine reptiles descended from egg-laying terrestrial reptiles. One might imagine that these marine reptiles returned to land to lay eggs and reproduce. However, from the preservation of embryos within ichthyosaurs it has long been

Figure 3.4 Model of the Late Cretaceous pterosaur *Quetzalcoatlus northropi* in flight. This halfscale working model was designed and built by a team at Aerovironment, Inc., led by Paul MacCready, an aeronautical engineer. When alive this pterosaur had a wingspan of 11 m. The design of this robot addressed questions on how this pterosaur flew without an aerodynamic tail structure, and therefore how it achieved pitch stability and yaw control, conditions that allow controlled directed flight by a flying animal. AeroVironment, Inc. website: http://www.avinc.com/uas /adc/quetzalcoatlus/. Reproduced with permission. (*See insert for color representation.*)

known that they experienced live birth. A recently discovered plesiosaur fossil, seen in Fig. 3.5, with a preserved embryo within an adult, has shown that these marine reptiles also experienced live birth. This circumstantial evidence has a strong impact on how we understand the ecology of these important predators of Mesozoic seas. The preservation of other organisms inside fossil organisms can also indicate aspects of diet. Preservation of fully articulated birds with feathers has been a hallmark of the Early Cretaceous Jehol Biota from northeastern China. Because of the presence of preserved feathers and other features much has been learned from this fauna about evolution and ecology of early birds. In particular, a number of specimens of *Yanornis* from the Jehol Biota contain ingested fish remains (Fig. 3.6), indicating that this taxon was primarily piscivorous.

Ultimately the study of function and morphology of fossil organisms is wrapped together not only with the evolutionary history of the organism being examined, but also with an organism's evolutionary heritage of biomineralization and how the organism is built through development. Dolf Seilacher was a big proponent of this viewpoint, as one can see in his portrayal of the influences on morphodynamics as a tetrahedron encompassing function, phylogenetic tradition, fabrication, and environment, shown in Fig. 3.7. For instance the materials that an organism has evolved to allow it to fabricate a skeleton are extremely important. The developmental genetic programming which the organism uses to produce pattern formation and chemical regulation, for example, is also included in this part of morphodynamics. Related to this is the overall constraint of phylogenetic tradition, which is also a function of the genome. This component emphasizes that the overall body plans of different groups of organisms are very important in constraining what that organism can evolve as function. Another feature of the tetrahedron is effective environment and indeed the physical and biological components that can be determined from environments both from physical and chemical as well as paleoecological approaches are important for understanding the environmental constraints under which an organism evolved. Function includes the internal and external

(a)

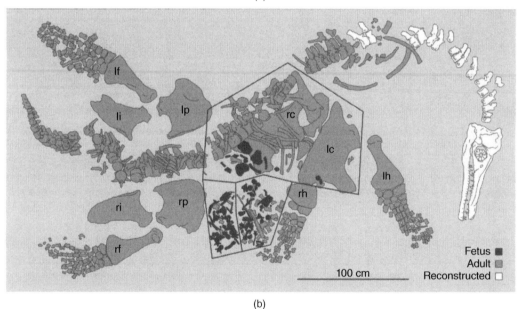

(b)

Figure 3.5 Skeleton (a) and interpretive drawing (b) of a fetus preserved within a pregnant adult Late Cretaceous plesiosaur *Polycotylus latippinus*. This is the first definitive evidence that plesiosaurs were viviparous. This evidence includes skeletal features showing that the smaller individual is a juvenile, taphonomic evidence that the juvenile was not consumed by the adult, articulation features of the juvenile skeleton indicating that it was within the adult at the time of burial, and skeletal features showing that both skeletons are *P. latippinus*. This relatively large single fetus indicates that plesiosaurs reproduced differently from other marine reptiles but does resemble the K-selected strategy of all modern marine mammals. The r-selected reproduction of other marine reptiles involves giving birth to several relatively small young, where parental investment is spread across these several young. The likely K-selected strategy of *P. latippinus* with parental investment concentrated in a small brood with large birth size may indicate that like modern marine mammals plesiosaurs were social and invested heavily in parental care. Labelled bones are listed in O'Keefe and Chiappe (2011). From O'Keefe and Chiappe (2011). Reproduced with permission from the American Association for the Advancement of Science. (*See insert for color representation.*)

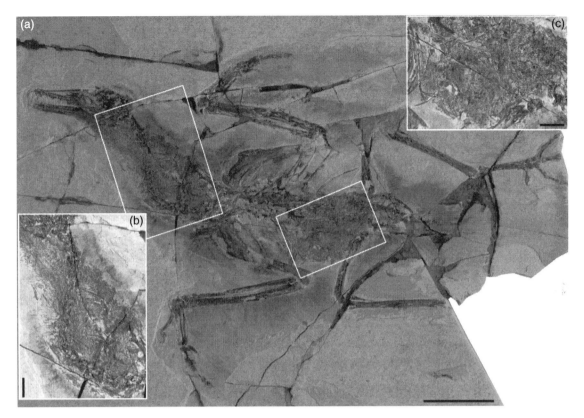

Figure 3.6 Specimen of the Early Cretaceous bird *Yanornis* (a) with ingested whole fish in the crop (b) and macerated fish bones in the ventriculus (c). This specimen of *Yanornis* is from the Lower Cretaceous Jehol Group (China) and is an early representative of the Ornithuromorpha, the lineage in which living birds are included. This and other specimens of *Yanornis* indicate that this taxon was a fish-eater, that it did not use its teeth to macerate fish before they entered the crop, and that fish were subsequently macerated in the ventriculus (gizzard or muscular stomach). Scale bars are 5 cm for (a), 1 cm for (b) and 1 mm for (c). From Zheng et al. (2014). Used under CC-BY-3.0 https://creativecommons.org/licenses/by/3.0/. (*See insert for color representation.*)

functions for organs and skeletal parts, and also a behavioral function for the morphology which we can see is another emphasis of morphodynamics.

Paleoecological models for paleoenvironmental reconstruction

Work done with Kathleen Campbell, Jennifer Schubert, and Mary Droser, outlined below, has focused on the process by which paleoecological models for paleoenvironmental reconstruction develop (Bottjer et al. 1995). To produce these models the various approaches to understanding the function of individual organisms, already discussed, can then be combined with additional information from facies analysis and geochemical processes. Paleoecological models for paleoenvironmental reconstruction proceed through a history of development that involves steady incorporation of new information, from modern and ancient environments and ecologies. All paleoecological models for paleoenvironmental reconstruction

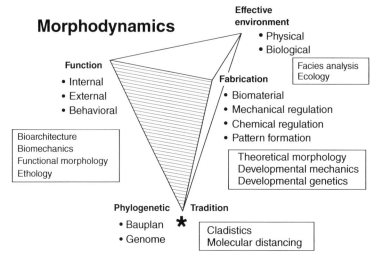

Figure 3.7 The conceptual framework of morphodynamics represented as a tetrahedron. The specific research fields for determining function include bioarchitecture, biomechanics, functional morphology, and ethology. The research fields utilized in determining phylogenetic tradition include cladistics and molecular approaches. The fields of theoretical morphology, developmental mechanics and developmental genetics provide information on fabrication. The effective environment is determined through facies analysis and ecological studies. From Briggs (2005). Reproduced with permission from the author.

have sets of paleontological, sedimentological, stratigraphic, and sometimes geochemical criteria that are used, in some cases loosely, in others fairly strictly, for interpretative decisions. To a large extent the level of rigor with which a paleoecological model is applied depends upon how formally it has been conceptualized, and how much agreement exists on the applicable features of the model to specific examples from the geological record. These models are usually designed to lead to a better understanding of depositional environments.

Through their history of use paleoecological models have developed in a variety of ways. New discoveries can lead to splitting-away of a subset of the phenomena originally thought to be explained by the model. This partitioning then may lead to the development of new paleoecological models for the newly delimited phenomena. New discoveries can also lead to the reevaluation of specific paleoecological criteria previously thought to indicate a particular environmental condition, leading to a refinement of the model. New discoveries may also demonstrate

the need for a general reevaluation of the model, or possibly, even abandonment of the model. In these ways, paleoecological models for paleoenvironmental interpretation transform and evolve just like any other scientific approaches to solving problems.

As an example of the success of actualism in interpretations of past ecological and environmental settings, a history of the scientific development of models to reconstruct ancient lens- to irregularly-shaped carbonate bodies with abundant macrofossils is illuminating. Paleontologists and sedimentary geologists have traditionally maintained a high level of interest in such carbonate bodies. Before the 1980s these fossiliferous carbonate bodies were usually interpreted to indicate deposition in shallow-water marine environments such as reef settings. For sedimentologists, this high level of interest has been generated for practical reasons – reef carbonates are typically reservoir rocks for petroleum. And, for the paleoecologist, the geological history of reef ecology has also attracted a significant amount of attention, because these diverse, dynamic communities show spectacular trends in evolution and extinction (e.g., Fig. 12.17).

In the modern, most scleractinian corals have a symbiosis with photosymbiotic algae termed zooxanthellae that allows protection for these microbes and greater growth rates for the corals. Because modern reef growth and development are linked directly to photosynthetic organisms that require a photic zone habitat, the predilection for an actualistic interpretation that such carbonate features were deposited in relatively shallow water has been compelling. Until the 1980s, perhaps the best documented example of how such straightforward actualistic approaches can lead to incorrect interpretations is the occurrence of azooxanthellate scleractinian corals that produce mounds or build-ups with constructional frameworks in deep-water environments, which in the stratigraphic record are potentially confused with shallow-water reefs. So, in response to the anomaly represented by the discovery in modern environments of deep-water mounds and buildups, the actualistic understanding of such carbonate bodies underwent revision, so that such bodies could be interpreted as potentially deposited in either shallow or deep water (Fig. 9.1).

Further development of actualistic paleoecological models for determining paleoenvironments of these ancient limestone deposits has been similarly incremental, as more has been learned about modern environments that can foster deposition of lens- to mound-shaped carbonate bodies. And, in particular, research since the beginning of the 1980s in the broad study of such deposits has led to the realization that many carbonate bodies which were formerly interpreted as reef and associated shallow-water deposits may in fact be the fossilized remains of deeper-water cold seeps.

As an example which illustrates this trend, near Pueblo, Colorado (USA) numerous limestone bodies occur within the Upper Cretaceous (Campanian) Pierre Shale, as shown in Fig. 3.8. These carbonates are more resistant than the shales so that in surface outcrops they tend to erode in a topographically characteristic conical shape, called "Tepee Buttes." A typical Tepee Butte consists of a cylindrical,

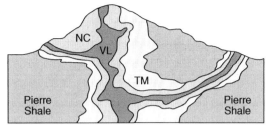

Figure 3.8 Cross-section of a typical Tepee Butte within the Pierre Shale. These are fossilized Cretaceous methane-seep ecosystems, and numerous examples of these mounds which may be as high as 20 m (hence the name "butte") preferentially weather-out near Pueblo, Colorado. The vuggy limestone (VL) marks the central vent. A coquina of the lucinid bivalve *Nymphalucina* typically surrounds the vent (NC). A thrombolitic micrite (TM) then drapes these central facies. In the process of metabolizing the venting methane as well as associated sulfates microbes increase the carbon dioxide concentration leading to precipitation of carbonate minerals. Chemosymbiotic bacteria are inferred to have lived within the tissues of *Nympholucina* as well as tube worms found in these deposits. From Shapiro and Fricke (2002). Reproduced with permission from the Geological Society of America.

vertical core with vuggy carbonate and abundant, articulated specimens of the lucinid bivalve *Nymphalucina occidentalis*. These occurrences were earlier interpreted to indicate biotic colonization by these bivalves in lagoonal grass beds. The actualistic model used for this interpretation included a modern analog of marine grass banks (which also contain lucinid bivalves) that currently exist in the US Virgin Islands.

At the same time that the Cretaceous Tepee Buttes were being diagnosed as having a shallow-marine grass bank origin, announcement was made of the discovery of modern hydrothermal vent faunas in the deep sea. Unexpectedly, large macroinvertebrates (molluscs, tube worms) were found flourishing at fluid venting sites along oceanic spreading centers, in marked contrast to the otherwise typical deep-sea faunas in the surrounding environment. Subsequently, many of these larger macroinvertebrates were found to contain chemosymbiotic bacteria that release the energy locked-up in the reduced, sulfide- or methane-rich vent fluids to generate metabolites for the large hosts. In particular modern lucinid bivalves have been found to be chemosymbiotic. Hence, with the discovery of chemosynthetically based ecosystems at hydrothermal vents, and later at hydrocarbon cold seeps and elsewhere, a new actualistic mechanism could be invoked to explain dense macrofossil associations in various deeper-water, non-photic-zone ancient marine settings, as well as shallower-water paleoenvironments.

Moreover, hydrothermal vents and cold seeps by their nature also provide point sources of fluids to the overlying depositional environments. For example, closely associated with hydrocarbon seeps are isolated anomalous carbonates precipitated at the seafloor when methane-rich fluids contact sea water. Therefore, an additional mechanism that leads to *in situ* precipitation of carbonate lenses and mounds in deep-water marine depositional settings was then available for application in an actualistic way to interpretations of ancient strata. And subsequent palaeoecological and geochemical work on the Tepee Buttes (Fig. 3.8), with their presumably chemosymbiotic lucinid bivalve fauna, has indeed verified their origin as submarine springs deposited in a deeper-water (several tens to hundreds of meters) terrigenous seaway.

Paleoecology and paleoclimate

Palaeoecological models have also been developed to determine aspects of paleoclimate using a variety of terrestrial fossils. Fossil leaves have been a particularly important source of information on ancient climate. For example, stomata are pores on leaf surfaces through which plants exchange CO_2, water vapor, and other components with the atmosphere. Through observations of modern plants in different CO_2 concentrations a general relationship has been observed where the number of stomata decreases with increasing CO_2 concentrations, and likewise increases with decreasing CO_2 concentrations. A

Stomatal Index has been devised where Stomatal Index = (number of stomata/number of stomata + number of epidermal cells) × 100. Therefore, there is an inverse relationship between leaf stomatal indices (stomatal density) and the partial pressure of atmospheric CO_2. The development of stomata on leaves varies between different plant taxa, although it appears to be consistent within taxa. Thus, it appears best to use the stomatal index where there is a modern representative that can be studied with results extended into the past by means of a uniformitarian approach. Greg Retallack has shown that the leaves of the *Ginkgo* tree can be used in this way. *Ginkgo* has a fossil record at least back to the Late Triassic. To extend the record further back into the Paleozoic, plants with a Mesozoic and Paleozoic record that co-occur in the Late Triassic with *Ginkgo* leaves and have the same stomatal indices as the

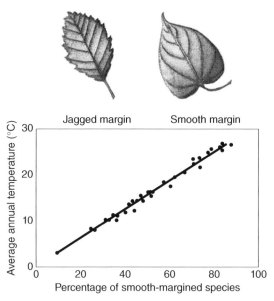

Figure 3.9 Relationship between leaf margin morphology and temperature as a paleothermometer. These analyses are made from assemblages of leaves collected from a single sample horizon, and have provided significant information on temperatures for the last 100 million years. Smooth-margined leaves are more characteristic of warmer temperatures. From Stanley (2008). Reproduced with permission from W.H. Freeman.

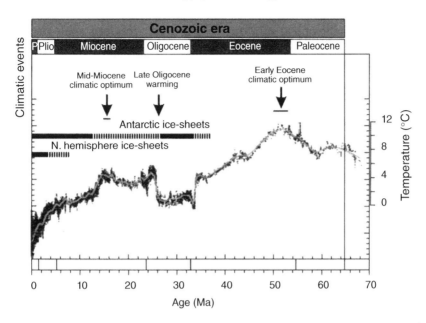

Figure 3.10 Cenozoic paleotemperatures determined from stable oxygen isotope variations in foraminifera shells. Before 34 million years ago the data is a record of deep ocean temperatures. After 34 million years ago, continental ice sheets developed, so that the signal is a mixture of temperature and the effects of ice volume. Note the late Paleocene and early Eocene warm intervals as well as the late Oligocene warming and mid-Miocene climatic optimum which overlie a broad temperature decline as the Earth has progressed from a Greenhouse state in the Eocene to its current Icehouse state (see also Fig. 1.9). From Beerling (2008). Reproduced with permission from Oxford University Press.

Ginkgo leaves have been used. Thus, Retallack has shown that in one example fossil pteridosperm leaves which co-occur with ginkgo leaves provide stomatal index data also for the Permian.

Fossil leaves have also been utilized in other ways to determine paleoclimate. Fossil angiosperm leaf margin analysis is a univariate method that allows the determination of paleotemperatures when fossil leaves were alive. The first requirement is to obtain a collection of leaf fossils from a site that represents a large number of tree species. As shown in Fig. 3.9 one then determines the percentage of the species that have leaves with smooth or entire margins, as opposed to toothed, or jagged margins. This number – the percentage of smooth-edged leaves – goes into an equation that gives the average annual temperature (AAT) in Celsius of the given time and place: AAT = (0.3006 × percent smooth)

+ 1.141. This method works because teeth allow leaves to begin photosynthesis early in the spring, an advantage in climates with short growing seasons. On the other hand, teeth allow a loss of water vapor, a disadvantage in a warm climate. Thus, one sees a high percentage of smooth-edge species in warm regions, which has been observed in living forests around the world. Further refinement of paleoclimate determinations using leaf morphology through multivariate approaches, pioneered by Jack Wolfe with the development of CLAMP (Climate Leaf Analysis Multivariate Program), have reinforced the utility of fossil leaves for understanding ancient climate.

In water the proportion of the stable isotopes of oxygen, ^{18}O and ^{16}O, changes with temperature. For marine organisms with shells made of minerals that have oxygen in them, such as calcium carbonate or

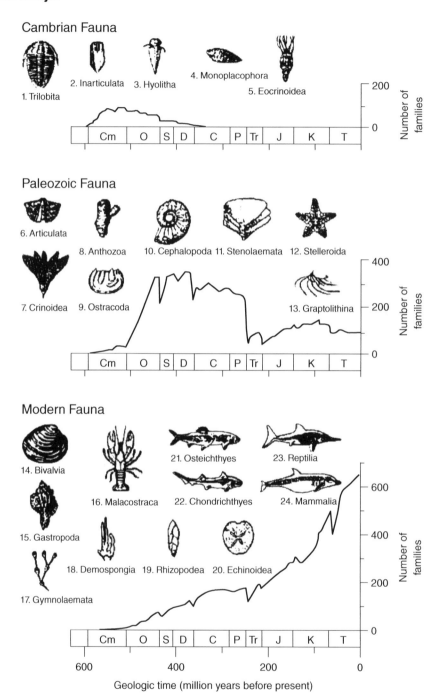

Figure 3.11 Familial biodiversity of the three Phanerozoic marine evolutionary faunas, as determined by Sepkoski through factor analysis of his marine Phanerozoic biodiversity data base. Each consists of broad sets of taxa that were globally dominant through long periods of geological time, with characteristic taxa for each fauna schematically displayed. The Cambrian Fauna includes many organisms characteristic of the Cambrian Explosion. The Paleozoic Fauna consists of organisms that characterize the Great Ordovician Biodiversification Event (GOBE). The Paleozoic Fauna was significantly affected by the end-Permian mass extinction, which led to dominance by the Modern Fauna in the post-Paleozoic. From Foote and Miller (2007). Reproduced with permission from W.H. Freeman.

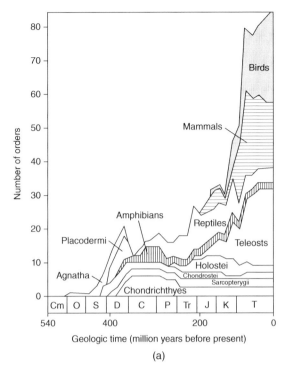

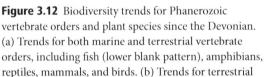

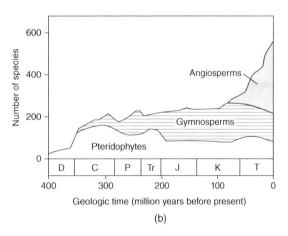

(a)

(b)

Figure 3.12 Biodiversity trends for Phanerozoic vertebrate orders and plant species since the Devonian. (a) Trends for both marine and terrestrial vertebrate orders, including fish (lower blank pattern), amphibians, reptiles, mammals, and birds. (b) Trends for terrestrial plants, including pteridophytes (vascular plants that reproduce by spores), gymnosperms, and angiosperms (more recent studies show an earlier angiosperm history beginning in the Late Jurassic). From Foote and Miller (2007). Reproduced with permission from W.H. Freeman.

calcium phosphate, the proportion of ^{18}O to ^{16}O can be determined to understand the temperature of the seawater at which the shell was precipitated. An increase in the ratio of $^{18}O{:}^{16}O$ can indicate that the temperature of precipitation for the skeleton was cooling. During a world with significant polar ice (Icehouse World), complicating factors are the amount of water that is locked up in ice, as the lighter ^{16}O is preferentially evaporated to eventually form ice, thus leaving the ocean with a higher ^{18}O. During a world free of significant polar ice (Greenhouse World), temperature is largely the controlling factor. The shells of foraminifera, made of calcium carbonate, are commonly used to produce paleotemperature records. A well-known example is the oxygen isotope record for the Cenozoic,

made from foraminifera, and the implications that it has for paleotemperature, as shown in Fig. 3.10. Conodonts, which are microfossils made of calcium phosphate that are the jaw elements of extinct eel-like chordates, are commonly also used for oxygen isotope paleothermometry, for their Cambrian through Triassic range.

Evolutionary paleoecology

Data on fossil occurrence through the stratigraphic record are painstakingly determined through detailed studies of stratigraphic sections, and published in a variety of scientific papers.

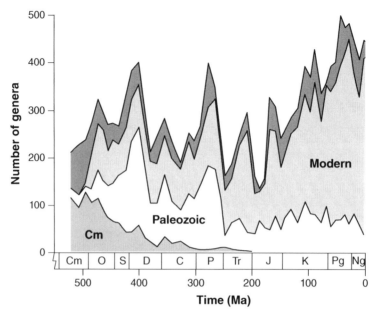

Figure 3.13 Phanerozoic biodiversity curve showing the three Phanerozoic marine evolutionary faunas, as determined by Alroy from a sampling standardized diversity curve generated from generic data in the Paleobiology Data base. The unlabeled area represents groups not assigned to one of the evolutionary faunas; Cm is Cambrian Fauna. From Alroy (2010). Reproduced with permission from the American Association for the Advancement of Science.

Paleontologists have been assembling these data in a usable fashion for the past 150 years. The ability to determine broad trends in the fossil record requires compilation of these data by combing through this century and a half of paleontologic and geologic literature. Building on the work done for the Treatise on Invertebrate Paleontology, first directed and edited by Raymond C. Moore, this sort of effort was pioneered by J. John Sepkoski, Jr., who spent much of the 1970s and 1980s compiling Phanerozoic fossil data by hand before the days of the personal computer. This work resulted in the famous "Sepkoski curve" of Phanerozoic marine biodiversity, considered to be one of the most widely used figures in paleontology over the past 30 years, and providing much stimulus for research of the fossil record. Paleoecological trends in the Phanerozoic were also elucidated by Sepkoski, through his statistical determination of the "three great evolutionary faunas," each with a different ecology and impact upon the biosphere, as shown

in Fig. 3.11. Similar biodiversity compilations have also been made for the vertebrate and plant records, as shown in Fig. 3.12. The efforts of Sepkoski have been continued with the development of the Paleobiology Database and Fossilworks, primarily through the work of John Alroy. This work has led to new versions of a Phanerozoic biodiversity curve, which also depicts the three great evolutionary faunas of the Phanerozoic, as shown in Fig. 3.13. In an allied realm, through the work of Eric Flügel and Wolfgang Kiessling, a fossil reefs database for the Phanerozoic has been developed. All of these efforts are part of the move of paleontology into the bioinformatics age and the enormous utility that such databases can have towards framing new directions of research.

Once we know where organisms lived, through depositional and palaeoecological models, and their life habits, through functional morphology analysis, we can then use this information to quantify how ecological occupation has changed through time,

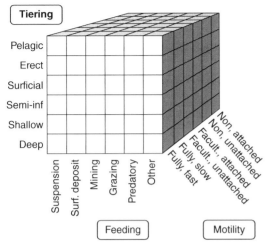

Figure 3.14 Theoretical ecospace use cube. Tiering indicates where the organism lives above and below the seafloor, and includes pelagic (in the water column), erect (benthic, extending into the overlying seawater), surficial (benthic, not extending significantly upward), semi-infaunal (partly infaunal and partly exposed), shallow (infaunal, living in the top ~5 cm of sediment), and deep (infaunal, living more than ~5 cm deep in the sediment). Motility level includes fully, fast (regularly moving, unencumbered); fully, slow (regularly moving but with a strong bond to the seafloor); facultative, unattached (moving only when necessary, free-lying); facultative, attached (moving only when necessary, attached); nonmotile, unattached (not capable of movement, free lying), nonmotile, attached (not capable of movement, attached). Feeding mechanisms include suspension (capturing food particles from the water), surface deposit (capturing loose particles from a substrate), mining (recovering buried food), grazing (scraping or nibbling food from a substrate), predatory (capturing prey capable of resistance), and other (e.g., photo-or chemosymbiosis, parasites). From Bush et al. (2007). Reproduced with permission from Cambridge University Press.

and gain a better understanding of the interplay between macroevolution and ecology. Such studies, initially pioneered by Richard Bambach, later joined by Andrew Bush, have defined how organisms partition ecospace, as shown in Fig. 3.14.

Summary

These various aspects of the science of paleoecology continue to receive intense interest and study. For example, in paleoenvironmental reconstruction, the utility of fossil leaf morphology towards paleoclimate analysis continues to become increasingly sophisticated (e.g., Peppe et al., 2011). Likewise, the full range of questions that can be posed by evolutionary paleoecology is under pursuit, starting with the fundamental question of the nature of the three great evolutionary faunas (e.g., Alroy, 2004). Through this continued refinement studies determining the interplay between environment, ecology, and evolution from the stratigraphic and fossil record have grown increasingly sophisticated. Thus paleoecology shows great promise to further play an integral role in understanding how life has evolved on Earth and how we might understand and manage future environmental and ecological change.

References

Alroy, J. 2004. Are Sepkoski's evolutionary faunas dynamically coherent? *Evolutionary Ecology Research* 6, 1–32.

Alroy, J. 2010. The shifting balance of diversity among major marine animal groups. *Science* 329, 1191–1194.

Beerling, D. 2008. *The Emerald Planet: How Plants Changed Earth's History*, Oxford University Press.

Bottjer, D.J., Campbell, K.A., Schubert, J.K. & Droser, M.L. 1995. Palaeoecological models, non-uniformitarianism, and tracking the changing ecology of the past. *In* Bosence, D. & Allison, P. (eds.), *Marine Palaeoenvironmental Analysis from Fossils*. Geological Society of London Special Publication, pp. 7–26.

Brenchley, P.J. & Harper, D.A.T. 1998. *Palaeoecology: Ecosystems, Environments and Evolution*. Chapman & Hall, London.

Briggs, D.E.G. 2005. Seilacher on the science of form and function. *In* Briggs, D.E.G. (ed.), *Evolving Form and Function: Fossils and Development*. Special Publication of the Peabody Museum of Natural History, Yale University, pp. 3–24.

Bush, A.M., Bambach, R.K. & Daley, G.M. 2007. Changes in theoretical ecospace utilization in marine fossil assemblages between the mid-Paleozoic and late Cenozoic. *Paleobiology* 33, 76–97.

Foote, M. & Miller, A.I. 2007. *Principles of Paleontology*, 3ʳᵈ *Edition*, W.H. Freeman, New York.

O'Keefe, F.R. & Chiappe, L.M. 2011. Viviparity and K-selected life history in a Mesozoic marine plesiosaur (Reptilia, Sauropterygia). *Science* 333, 870–873.

Peppe, D.J., Royer, D.L., Cariglino, B., Oliver, S.Y., Newman, S., Leight, E., Enikolopov, G., Fernandez-Burgos, M., Herrera, F., Adams, J.M., Correa, E., Currano, E.D., Erickson, J.M., Hinojosa, L.F., Hoganson, J.W., Iglesias, A., Jaramillo, C.A., Johnson, K.R., Jordan, G.J., Kraft, N.J.B., Lovelock, E.C., Lusk, C.H., Niinemets, U., Penuelas, J., Rapson, G., Wing, S.L. & Wright, I.J. 2011. Sensitivity of leaf size and shape to climate: global patterns and paleoclimatic applications. *New Phytologist* 190, 724–739.

Shapiro, R. & Fricke, H. 2002. Tepee Buttes: Fossilized methane-seep ecosystems. *GSA Field Guides* 3, 94–101.

Stanley, S.M. 1970. Relation of shell form to life habits of the Bivalvia (Mollusca). *Geological Society of America Memoir* 125, 1–296.

Stanley, S.M. 2008. *Earth System History*, W.H. Freeman, New York.

Zheng, X., O'Connor, J.K., Huchzermeyer, F., Wang, X., Wang, Y., Zhang, X. & Zhou, Z. 2014. New specimens of *Yanornis* indicate a piscivorous diet and modern alimentary canal. *PLOS ONE* 9, e95036.

Additional reading

Allmon, W.D. & Bottjer, D.J. (eds.), 2001. *Evolutionary Paleoecology: The Ecological Context of Macroevolutionary Change*. Columbia University Press.

Dodd, J.R. & Stanton, R.J. 1990. *Paleoecology: Concepts and Applications*, 2ⁿᵈ *Edition*. Wiley-Interscience.

Ivany, L.C. & Huber, B.T. (eds.). 2013. Reconstructing Earth's Deep-Time Climate. The Paleontological Society Papers 18. The Paleontological Society.

Tevesz, M.J. & McCall, P.L. (eds.), 1983. *Biotic Interactions in Recent and Fossil Benthic Communities*. Plenum Publishing Corporation.

4 Taphonomy

Introduction

Taphonomy is one of the most important subjects of paleoecology. Understanding how fossil remains became part of the record is crucial to any study. Thus, what the fossil record can be used for is dependent upon our understanding of taphonomy. This goes not only for paleoecology but for all areas of paleontology. Taphonomy is the study of what happens to a microbe, animal, or plant after it dies. This process can be broken up into two phases: (1) what happens before final burial in sediment and (2) what happens between final burial in sediment and discovery by a human observer. Any of a large number of taphonomic processes can act to destroy the remains of an organism, so that it does not survive as a fossil which can give information. Preservation therefore is a relatively rare event, when we consider the millions of organisms that continually live on the Earth. Not many, because of taphonomic processes, make it into the fossil record. If they did, we would be sitting here on the surface of the Earth on a huge pile of leaves, bones, and shells.

Magnitude of taphonomic processes

How does this taphonomic filter distort our understanding of the fossil record? We can get some sort of understanding of this by updating an analysis first done by Raup and Stanley (1971) through estimating how many species of plants and animals exist in the present and how many fossil animals and plants have been discovered and described for the scientific world. Estimates of the number of animal and plant species on Earth right now range from 4 to 15 million. For our analysis, though, let's use an estimate of 10 million.

Then, we know that animals and plants have been on Earth for at least 600 million years. We can say that the average length of time for a species on Earth has been about 3 million years. So, using these numbers, there has been a complete turnover of species on Earth about 200 times. And, then the number of potential animal and plant species that may have lived on Earth is 10 million times 200 which is 2 billion possible species. But paleontologists have described only about 300,000 fossil species. So there should be really many more fossil species in the

Paleoecology: Past, Present and Future, First Edition. David J. Bottjer.
© 2016 John Wiley & Sons, Ltd. Published 2016 by John Wiley & Sons, Ltd.

record, but there aren't. Why not? Taphonomic processes.

It turns out that if an organism lives in marine environments, which seem to be more hospitable to preserving biological remains, and if it has a well-mineralized skeleton, its chance of preservation is much better than is apparent from the previous arguments. Examples of those marine groups with good skeletons are foraminifera, echinoderms, sponges, corals, molluscs (clams, snails, ammonoids), brachiopods, and some arthropods (trilobites). These groups have what is called a good fossil record. Paleontologists like to study groups with good fossil records.

A group that does not have a good fossil record is the insects. About 750,000 living species are insects. However, only about 20,000 fossil species have been recognized. The basic reason for this is that insects do not have a mineralized skeleton and they do not live in marine environments.

Normal preservation

Figure 4.1 shows the several processes of taphonomy which follow organic remains from the biosphere to the lithosphere. Various filters exist through the steps that lead to fossilization. Upon death in unusual circumstances, organic remains may be immediately buried out of the reach of processes that occur at the surface, and thus, they become part of the lithosphere almost immediately and subject to the processes of exceptional preservation, discussed in the following text. More typically, there is delayed burial in the filtering process that is associated with exposed remains including biological and chemical processes. In many environments, organic remains may be recycled from burial back to the surface numerous times before they finally become part of the lithosphere. Once organic remains are permanently in the lithosphere, they then undergo a variety of diagenetic (rock-forming) processes that continue to affect fossil remains until they are collected by a paleontologist. That's where paleoecological reconstruction begins. However, the

quality of paleoecological reconstruction is strongly affected by our understanding of how the original biological information has survived through these various taphonomic filters.

One of the reasons that marine environments are generally better for fossil preservation is that terrestrial environments are generally subject to more erosional processes than many marine environments. Thus, in terrestrial settings, there is more of a possibility of erosion and transport of organic remains, with resultant destruction, than in a marine environment.

Within marine environments, the chance of preservation is affected by the original life mode of the organism. Infaunal organisms, which live below the seafloor, are commonly already buried in the sediment, while epifaunal organisms, which live on the seafloor, accumulate on the sediment before burial. Organisms which float or swim in the water column travel through some portion of the water column before arriving on the seafloor and subsequent burial.

The normal processes of fossilization begin with biological destruction, which is the first step essential to preservation. The effects of biological destruction vary depending upon life mode as different biological processes are in effect in different environments. In marine environments, the skeletons of most organisms are made of calcium carbonate. Most of the marine fossil record is composed of originally biomineralized skeletons that have commonly been disarticulated. A variety of other organisms like to bore into these calcium carbonate skeletons, particularly after the animal has died. They bore into skeletons until all that is commonly left is a pile of chips.

Second, there is mechanical destruction. Many marine environments, particularly nearshore ones, are subject to significant wave and current action, which moves skeletons around a lot and grinds them up to bits. We have all been beachcombing and seen this process in action – all of those shells thrown up on the beach are subject to taphonomic processes of mechanical destruction. Third, there is chemical destruction. A skeleton can be dissolved

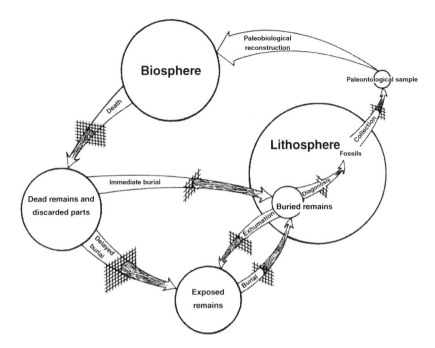

Figure 4.1 Flow diagram of organic remains as they progress from death to becoming a fossil to collection by the paleontologist. The effectiveness of the various filters in taphonomic processes allows different degrees of paleobiological and paleoecological reconstruction from each paleontological sample. In paleoecological studies that involve statistical comparisons of numerous samples, it is important that the samples are taphonomically comparable. From Behrensmeyer and Kidwell (1985). Reproduced with permission from Cambridge University Press.

in some marine environments where seawater is not saturated in dissolved calcium carbonate. Probably, the most common form of chemical destruction occurs after the fossil has been buried, when diagenesis is operating.

Under typical taphonomic conditions where oxygen is present in marine settings, scavenging and microbial decay rapidly remove soft tissues from mineralized skeletal elements such as shells and bones. These elements are also subject to scattering by carnivores and scavengers, degradation by agents such as boring microorganisms, chemical dissolution, and physical erosion by waves and currents. Thus, biological remains are typically destroyed before they can be buried by sediment.

However, a small proportion of organic remains do become buried below the seafloor. If the sediment pore waters are undersaturated in dissolved calcium carbonate (the mineral of which most shells are made) or calcium phosphate (of which bones are made), chemical dissolution will occur. If they are not dissolved, continued deposition of sediment can bury organic remains to the point where they are no longer in the taphonomically active zone (TAZ) and become immune to reexposure by erosion and damage by organisms that burrow through the sediment surface. It is these biological, chemical, and sedimentary processes that almost all fossils must pass through in order to become preserved.

In the marine realm, taphonomic filters at the sediment–water interface are varied, and several depictions of these filters and processes are illustrated in Figs. 4.2 and 4.3. Figure 4.2 shows living epifaunal organisms on the sediment–water

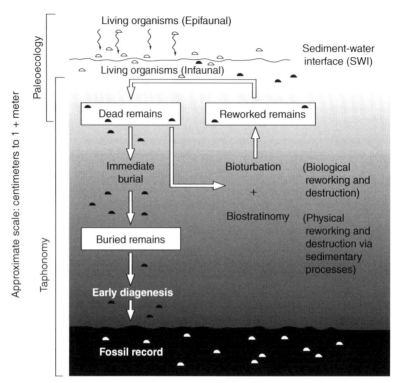

Figure 4.2 Progression for benthic organisms of taphonomic processes from living to dead to buried remains to final incorporation into the fossil record. Organic remains may experience multiple episodes of reworking through processes of bioturbation and biostratinomy. From Martin (1999). Reproduced with permission from Cambridge University Press.

interface as well as infaunal organisms below the sediment–water interface. Progression down the sediment column enters the realm of taphonomy. Figure 4.2 illustrates the dichotomy that can occur where remains are immediately buried and then subject to diagenesis versus those that are reworked. Reworked remains can be subject to a variety of processes including bioturbation. The biological reworking as well as destruction of organic remains can also be accompanied by physical reworking and destruction via nonbiological sedimentary processes. In another depiction of these processes, Fig. 4.3 shows additional geochemical processes for remains with calcium carbonate skeletons, including those of seafloor dissolution and diagenesis, and also depicts the TAZ as the zone of reworking. In this depiction, the TAZ is governed by the boundary between more oxygenated or oxic environments

and underlying anoxic sediments. The TAZ is where organic remains are not only physically bioturbated but also where the sediments are infused with oxygen from surface waters through a variety of activities by bioturbators, termed bioirrigation. In this depiction, the zone below the TAZ is where shallow burial diagenesis commences. Various burial diagenetic processes combine to produce a fossil record of shells and molds (Fig. 4.3).

Figure 4.4 depicts an analysis done with data from the Paleobiology Database that shows the change in taxonomic composition of the fossil record for marine environments through the Phanerozoic. This record is not the product of exceptional preservational processes, discussed in the following text, but rather was produced by normal or typical taphonomic processes, as are shown in Figs. 4.1–4.3. This normally preserved fossil record represents the

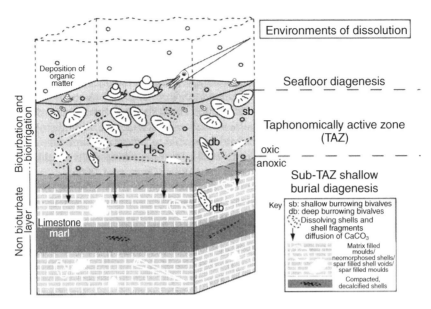

Figure 4.3 Progression of taphonomic processes for benthic organisms with calcium carbonate skeletons emphasizing environments of dissolution. Shells can be destroyed on the seafloor surface by bioerosion, physical processes, and dissolution. In the TAZ, increased acidity produced by decay of organic material and reoxidation, producing H_2S, can cause dissolution of high-Mg and aragonitic shells. Under these conditions, only minor amounts of aragonitic shells, such as deep-burrowing bivalves, would survive into the sub-TAZ diagenetic zone. Carbonate from dissolution of shells contributes to preferential early cementation of limestones allowing better three-dimensional preservation of shells than in shale beds, where fossils are more compacted. From Cherns et al. (2011). Reproduced with permission from Springer Science and Business Media.

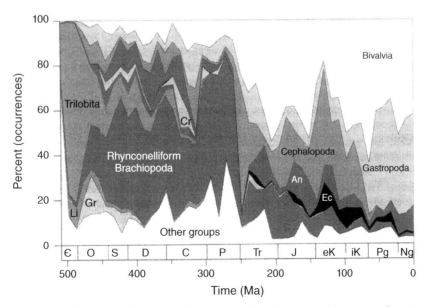

Figure 4.4 Variations in the Phanerozoic frequency of occurrence for major taxonomic groups of marine macrofauna. Data is from the Paleobiology Database that was downloaded 9/4/2007. Li, Lingulida; Gr, Graptolithina; Cr, Crinoidea; An, Anthozoa; Ec, Echinoidea. From Hendy (2011). Reproduced with permission from Springer Science and Business Media.

bulk of the data in the Paleobiology Database and presents the most extensive data on what we can learn about marine paleoecology.

Plants in terrestrial environments also undergo biological, mechanical, and chemical destruction. A unique quality of terrestrial plants is that they disarticulate and produce many parts (Fig. 4.5), which then are given different taxonomic names and which may travel various distances from where they grew. Pollen is the most easily transported, with pollen transported great distances in rivers or by wind. Leaf transport depends upon the height of the tree, whereas stumps are likely to be preserved *in situ*. Plant remains deposited within waterlogged sediments have the best potential to avoid decay, such as in wetland environments that produce peats and coal. Whole forests can be buried by volcanic eruptions, which can also lead to silicification of plant parts (Fig. 4.6). Resistance to degradation by plant parts varies, with flowers being among the least resistant. Figure 4.7 shows a schematic of terrestrial environments in which plant remains are preserved. Depending upon various processes of transportation and burial, different sorts of vegetation from different environments are preserved in different percentages.

The preservation of vertebrates in terrestrial environments also depends upon a variety of

Figure 4.5 A Permian assemblage exhibiting parts of a variety of plants. This preserves an autochthonous Late Permian forest-floor litter in which leaves of the gymnosperm *Glossopteris* (fossil at top of lower right quadrant) are preserved with horsetail (sphenopsid) ground cover which includes *Trizygia* (leaves in upper center and upper right quadrant; a cone in left center of lower left quadrant) and *Phyllotheca* (two foliar whorls showing long, narrow leaves that fit into a funnel-shaped sheath in left center of upper left quadrant). Such litter horizons, from the Upper Permian Balfour Formation, are an uncommon feature of the sequence transitioning the Permian–Triassic boundary, as defined by vertebrate biostratigraphy, in the Karoo Basin, South Africa. Locality and additional paleoenvironmental information can be found in Prevec et al. (2010) and Gastaldo et al. (2014). Photograph by Robert A. Gastaldo. Reproduced with permission. (*See insert for color representation.*)

Figure 4.6 Silicification of fossil forests buried by volcanic eruptions. On the western side of the island of Lesvos, Greece, volcanogenic debris flows and lahars preserve autochthonous Miocene forests, where erect, silicified trunks of *Taxodioxylon* (*Sequoia*) are preserved to a height of 7.1 m. The 350 m-thick Miocene volcaniclastic sequence, known as the Sigri Pyroclastic Formation, erupted from the Barossa stratovolcano and buried at least two major forest horizons. From Mpali Alonia Park in the Lesvos Petrified Forest GeoPark, Robert A. Gastaldo for scale. Photograph by Robert A. Gastaldo. Reproduced with permission. (*See insert for color representation.*)

factors ranging from climate and environment to the chemistry of sediments in which bones are deposited (Fig. 4.8). Vertebrates easily disarticulate into the component bones in terrestrial environments. In land environments, scavengers (vultures, hyenas, etc.) scatter and destroy bones. On land, streams and rivers transport bones and fracture and grind them up. Complete skeletons are most commonly preserved in settings where the organism

was buried quickly after death (or sometimes simultaneously with death). A broad schematic of the sorts of effects of taphonomic processes in the deposition of vertebrate fossils is depicted in Fig. 4.8. Here, processes on the macroscale at the global level that are due to climate are particularly important. On the mesoscale, the suite of environments in which animals live is similarly important for the taphonomic processes which ensue. Microscale processes as well determine how bones become part of the lithosphere. This hierarchy of processes also plays out for the remains of organisms in marine environments as discussed earlier.

Exceptional preservation

Organic remains that typically progress straight into the lithosphere without much of a stop in the TAZ, in marine or terrestrial environments, can be subject to taphonomic processes leading to exceptional fossil preservation. This type of fossil deposit is truly rare. These rare fossil deposits preserve the remains of soft tissues or the articulated nature of skeletal elements. Such deposits with exceptional preservation have been known for centuries, and because they have long been popular with the public, it is not unusual to find museums associated with them.

These deposits of exceptional fossil preservation were typically considered to be scientific curiosities, where, because of unusual preservation, paleontologists could decipher usually unobtainable aspects of organism morphology, behavior, function, and evolution. They have long had significance as the place where soft tissues fundamental to the understanding of the evolution of life, such as the feathers of *Archaeopteryx* or of dinosaurs, are found.

However, it is only over the past few decades that a concerted effort has been made to understand the overall significance of these rare and remarkable fossil deposits. Dolf Seilacher first popularized the general use of the word "fossil-lagerstätten" (singular, "lagerstätte"). In German, a lagerstätte is a geological deposit of economic interest, and a

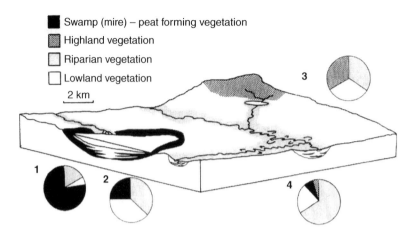

Figure 4.7 Block diagram showing a fluvial sedimentary environment with a superimposed vegetational mosaic and resulting composition of fossil leaf assemblages from different environments. This fluvial landscape of moderate relief has a meandering river crossing that includes numerous oxbow ponds. A small lake exists just downstream of the highland portion of the river. In the lowlands, the river feeds into a basin that includes a lake that fluctuates with swamp conditions. The flora includes a mostly swamp flora (including water's edge and mire communities that form deposits of peat), a dry lowland flora (including river margin or riparian components and an interfluvial forest), as well as a highland forest flora. The hypothetical composition of leaf assemblages in several sedimentary facies from these environments is shown in the pie charts. A swamp-coal facies (1) includes mostly flora from wetland environments as well as a component of lowland vegetation. Alternating with the swamp-coal facies are lake deposits (2) which contain a much greater component of lowland vegetation, due to fluvial transport. The highland lake (3) includes highland vegetation as well as components of lowland vegetation. The oxbow pond facies (4) is dominated by lowland flora including a large riparian component. Thus, the overall floristic composition of each facies is strongly affected by selective taphonomic processes. From Greenwood (1991). Reproduced with permission from John Wiley & Sons.

fossil-lagerstätte is a deposit containing fossils that are so exceptionally preserved or abundant that it warrants special exploitation, if only for scientific purposes. There are two types of fossil-lagerstätten: conservation lagerstätten, which contain exceptionally preserved fossils such as the impressions of soft parts of organisms, and concentration lagerstätten, which contain an abundance of fossils that are typically not exceptionally preserved. The term "fossil-lagerstätten" is now so widely accepted that paleontologists commonly shorten it to lagerstätten. Significant strides have been made in the past three decades, driven in large part by the need to understand the nature and preservation of early metazoan animals preserved in such deposits. Of special importance is that unique paleoecological information is available from these special sites.

Major types of lagerstätten are produced by a variety of processes, including stagnation, obrution, and microbial sealing in marine environments and in terrestrial environments burial by floods and volcanic deposits, as well as trapping. Three of the major processes that lead to the formation of marine conservation lagerstätten are included in the triangular diagram shown in Fig. 4.9. Figure 4.9 also shows a number of different marine conservation lagerstätten that are famous in the world of paleontology and where they stand in the field depicted by this triangular diagram.

Stagnation is a marine environmental condition where the body of water overlying the seafloor is

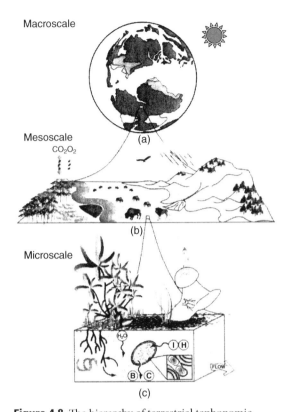

Macroscale

Mesoscale
CO_2O_2

Microscale

Figure 4.8 The hierarchy of terrestrial taphonomic processes and controls on the vertebrate fossil record. (a) Macroscale features include distribution of continental plates, sea-level variations, ocean circulation variations, differences in atmospheric composition and circulation, variability of intensity and distribution of solar radiation on the surface, and biome distribution. (b) Mesoscale landscape characteristics include variations in local weather patterns as well as differences in species population dynamics, biogeochemical cycles, predation/death, and scavenging of remains. (c) Microscale features include variations in soft tissue decay, bone exposure, desiccation and cracking from solar radiation, as well as invertebrate utilization, bioturbation, nutrient use and organic acid release by plant roots, leaching of bone mineral (B) and collagen (C), incorporation of exogenous ions (I) and humics (H) into bone matrix, bacterial and fungal degradation (inset), diagenesis, and hydraulic flow of groundwater. From Noto (2011). Reproduced with permission from Springer Science and Business Media.

largely devoid of oxygen, or is even anoxic, and therefore inhibits scavengers. Thus, organisms as organic remains come to rest on the seafloor and are preserved as whole articulated skeletons where soft tissues also have a good chance of being preserved as mineral replicas. These conditions are found in many marine low-oxygen basins but also can be found in lake deposits where a bottom low-oxygen layer had developed. A good example of stagnation deposits is the Jurassic Posidonia Shale in the Holzmaden area of southern Germany, which contains a spectacular assemblage of marine vertebrates and invertebrates, including many articulated marine reptiles and incredible specimens of crinoids. This shale was deposited during anoxic conditions and is typically laminated, although there are horizons with benthic fossils and bioturbation reflecting periods of intermittent oxygenation (Figs. 8.13 and 8.14). After death, animals deposited in anoxic conditions were not scavenged and thus remain articulated as fossils. The outline of soft tissues is sometimes preserved as a black film.

Obrution is a process where there is sudden burial by a layer of sediment such as occurs during storms. This process also occurs in terrestrial environments, as floods can rapidly bury organic remains and instantly put them into the lithosphere. In obrution, an organism is buried before it can be disturbed, and the quality of preservation may be very high (Fig. 4.10). Organisms (either alive or recently dead) are smothered and immediately buried by a pulse of sedimentation, such as a storm or turbidite. Once the remains are buried below the TAZ, they are protected from further disturbance.

Due to the lack of disturbance under both obrution and stagnation conditions, skeletal articulations will be maintained, but soft tissues will commonly still not be preserved. Soft tissue preservation is through the presence of unique geochemical conditions, commonly mediated by the presence of microbial activity, which leads either to the preservation of original organic material or to the early precipitation of pyrite, calcium carbonate, or phosphate, which produces a mineral replica of the soft tissues.

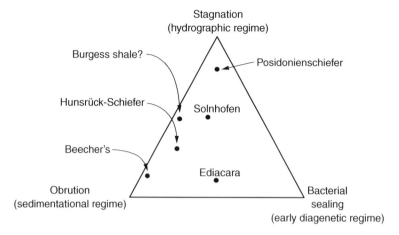

Figure 4.9 Some major marine conservation lagerstätten and the relative role in their preservation of obrution, stagnation, and bacterial sealing processes. The Ediacara biota from South Australia (but also of global extent) is of late Precambrian Ediacaran age and includes the oldest macroscopic animal fossils; the Burgess Shale, from British Columbia in Canada, is of Middle Cambrian age and exemplifies the faunas of the Cambrian explosion; the Ordovician Beecher's Trilobite Bed, from New York state in the United States, is well known for including pyritized trilobite soft tissues; the Hunsrück-Schiefer of Germany is of Devonian age and also includes extensive pyritized soft tissues; the Lower Jurassic Posidonienschiefer (Posidonia Shale; see also Figs. 8.13 and 8.14) from Germany includes spectacular specimens of marine reptiles and marine invertebrates; and the Upper Jurassic Solnhofen of Germany includes broad preservation of marine and terrestrial organisms with soft tissues including the bird *Archaeopteryx*. From Martin (1999). Reproduced with permission from Cambridge University Press.

A good example of exceptional fossil preservation by obrution is the spectacular Chengjiang fauna of Yunnan Province in southwestern China (Fig. 7.14). This early Cambrian lagerstätten has become so well known that it is now considered to be potentially more valuable toward understanding the Cambrian explosion than what we have learned from study of the Middle Cambrian Burgess Shale (Figs. 7.15 and 7.16). Figure 4.11 shows the importance of obrution and storm-generated currents to preserving the Chengjiang fauna where many fossils were originally soft-bodied organisms. Storm currents generated the deposition of sediment, and these event deposits sufficiently buried the early Cambrian organisms of the Chengjiang biota so that they passed directly into the lithosphere where they could then be preserved.

Bacterial sealing is a process whereby microbial mats are involved in covering the organic remains in an aqueous environment and aid in maintaining the integrity of the organic remains so that articulated skeletons are preserved and mineral replicas of soft tissues are also formed. Microbes are involved in preserving soft tissues, from providing depositional conditions that enhance preservation to producing mineral replicas of soft tissues during diagenesis. In aquatic environments where biofilms and mats can develop and there is an absence of disturbance by such processes as bioturbation, microbial mats are a key component for producing conservation lagerstätten.

The soft-bodied Ediacara biota forms a crucial piece of evidence for early macroscopic life on Earth. Preservation of these fossils associated with sandstones occurs in a different context from most soft-bodied preservation. Evidence has developed that, just as for carbonate substrates, Neoproterozoic siliciclastic seafloors were typically covered with

Figure 4.10 Crinoid assemblage from the Lower Mississippian Maynes Creek Formation, Le Grand, Iowa. Burial by storm deposition led to the instantaneous preservation of complete crinoid crowns, thus providing an "ecological snapshot" of this assemblage. Such preservation provides unparalleled resolution of morphological variability within specimens. Width of the darkest, common crinoid is 1 cm. Photograph by William I. Ausich. Reproduced with permission. (*See insert for color representation.*)

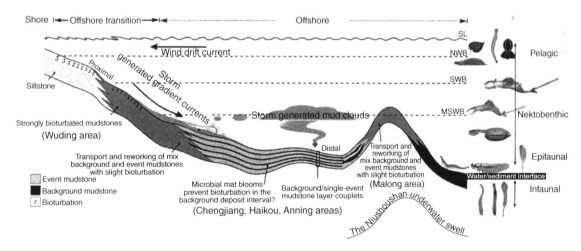

Figure 4.11 Depositional model for the early Cambrian Chengjiang biota in Yunnan, China. Onshore–offshore transect shows the distribution of important depositional processes for exceptional preservation of fossils, including storm-generated currents and presence of bioturbation and microbial mats. Facies with interbedded background mudstones and event mudstones are where the best specimens with soft tissue preservation are found, at localities in the Chengjiang, Haikou, and Anning areas. Schematic representations of organisms emblematic of the Chengjiang biota, and their interpreted ecology in this Early Cambrian marine environment, are displayed on right. SL is sea level, NWB is normal wave base, SWB is storm wave base, and MSWB is maximum storm wave base. From Zhao et al. (2012). Reproduced with permission from Elsevier.

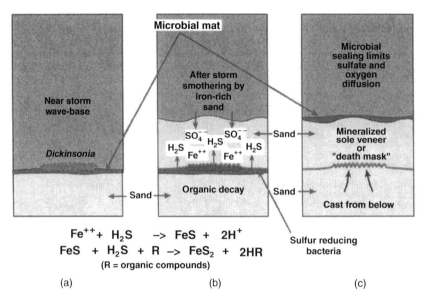

Figure 4.12 Microbial mats and preservation of the Ediacara biota. (a) A typical nearshore Ediacaran sandy seafloor covered by a microbial mat on which rests an individual of the Ediacaran animal *Dickinsonia*. (b) An event such as a storm has covered the seafloor with a layer of sand, and the *Dickinsonia* as well as the microbial mat on which it lived begins to decay, leading, in the presence of seawater sulfate, to the production of hydrogen sulfide which combines with the iron in the sediment to produce iron sulfide and, ultimately, with the inclusion of organic compounds, pyrite. (c) A microbial mat forms on the event bed, limiting sulfate and oxygen diffusion into the overlying seawater, while the pyrite forms a sole veneer or "death mask" which casts the *Dickinsonia* from below. From Droser et al. (2006). Reproduced with permission from Elsevier.

microbial mats. Jim Gehling has proposed that when event beds covered siliciclastic seafloors on which Ediacara biota organisms lived, the smothered microbial mats inhibited the vertical movement of pore fluids, hence promoting rapid cementation of a sole veneer in the overlying sand. In this way, the microbial mats may have acted as "death masks" for buried Ediacara organisms.

This process of microbial sealing for the Ediacara biota is illustrated in Fig. 4.12. In column A, the fossil *Dickinsonia* is shown living on the seafloor on top of a microbial mat. In column B, this bacterial sealing process is combined with obrution, and so there is deposition of a storm layer upon the *Dickinsonia,* which in column C is then covered by another microbial mat. This overlying microbial mat, through sealing the storm bed, leads to diagenetic processes which produce a mineral replica of *Dickinsonia* on the sole of the storm bed, which is the fossil that we see on outcrop as *Dickinsonia*. In Ediacara biota environments, this process would have been repeated many times. A variant on this process for Ediacara biota fossils and biogenic sedimentary structures is shown in Fig. 4.13. Here are shown frond-shaped organisms of the Ediacara biota with holdfasts that existed within or below a microbially bound sediment surface. During storm events, these fronds may have been dislodged or removed, while the holdfasts were variably dragged and stretched within the substrate. Subsequent burial by the storm layer led to preservation of the most distorted holdfasts as the Ediacaran biogenic sedimentary structure informally known as "mop" (Fig. 4.13).

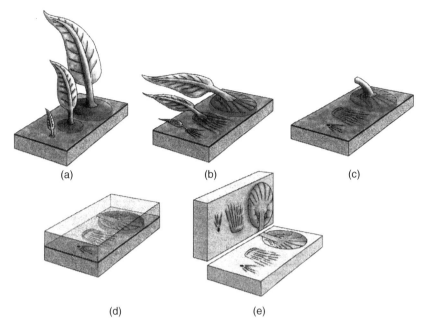

(a) (b) (c)

(d) (e)

Figure 4.13 Schematic illustration of hypothetical sequence of events in the formation of the biogenic sedimentary structure "mop" from the Ediacara Member of the Rawnsley Quartzite in South Australia. (a) *Charniodiscus*-like organisms with circular holdfasts attached within or beneath a nearshore microbially bound sandy substrate with stalks and fronds oriented vertically in the water column (three sizes depicted to illustrate the effect of size on mop formation). (b) An event including unidirectional currents flowing from right to left produces stress resulting in bunching, compression, and puckering of the holdfasts, including dragging, stretching, and shearing of the substrate by the smaller holdfasts. (c) The stress of the event current results in severing of the stalks, leaving some or none of the stalk. (d) The event causes deposition of a sand bed and compression of the remaining holdfast tissue and associated microbial and sedimentary structures, with subsequent lithification of the beds. (e) In the modern, a paleontologist separates the beds to reveal preservation of a puckered disk (right) and two smaller morphological variants of mop. From Tarhan et al. (2010). Reproduced with permission from the SEPM Society for Sedimentary Geology.

How are fossils preserved with all the parts attached? One of the best ways is to suddenly bury them. On land, this can be done with a flood or a volcanic ash flow. Akin to obrution are volcanic pyroclastic flows and ash falls that suddenly bury animals and plants, killing them and removing them from the TAZ. The oldest example of the Ediacara biota, from Mistaken Point, Newfoundland (Canada), occurs on bedding planes representing deepwater deposition, which were covered by ash from local eruptions, under which numerous fossils are preserved (Fig. 4.14). Forests can also be buried *in situ* and preserved by pyroclastic flows (Fig. 4.6). Perhaps the most famous example of this style of preservation is Pompeii, where molds of people buried in flows have been found.

The spectacular Florissant lagerstätten fossils were deposited in the Late Eocene in Colorado. Volcanic eruptions were common in this area, and pyroclastic flows like that which formed Pompeii occurred (Fig. 4.15). One flow inundated a redwood forest, and silica from the volcaniclastics later silicified the forest trunks, including stumps of large redwoods.

Figure 4.14 Ediacara biota fossils on the "E" surface from the Mistaken Point Formation, showing the stalked suspension feeder *Charniodiscus spinosus* (disk with attached frond below coin), and various rangeomorphs. This surface was preserved by volcanic ash that covered a living community after falling into the ocean. Opinions on the phylogenetic affinities of *Charniodiscus* are varied, and rangeomorphs are considered to be an extinct group that did not survive into the Phanerozoic. The fossils from Mistaken Point, Newfoundland, Canada, are considered to be part of the Avalon assemblage of the Ediacara biota, which make them the oldest Ediacara fossils, dated at 575–560 million years ago (see also Fig. 7.7). Coin is 24 mm. Additional information can be found in Narbonne et al. (2007). From Photograph by David J. Bottjer. (*See insert for color representation.*)

Deposition of volcanic sediments can contribute in other ways to exceptional preservation of fossils. Pyroclastic flows at Florissant dammed local river systems to produce lakes. Along with burial and preservation of forests, volcanic eruptions led to enrichment of lake water in silica leading to blooms of diatoms. Diatom masses on the surfaces of lakes incorporated plants and insects that were trapped in the mats. Ultimately, the mats sank to the lake bottom where the diatom layer and the incorporated plants and insects, for which the Florissant is most famous, were then preserved (Fig. 4.15).

In volcanic areas such as Yellowstone National Park (United States), there are surface hot springs which can also contain silica-rich waters. These waters can silicify adjacent organic remains. One of the best known examples of this process from the fossil record is the Lower Devonian Rhynie chert lagerstätten in Scotland, which contains silicified plants and animals that lived adjacent to hot springs and preserves an early terrestrial ecosystem (Fig. 11.1).

Various types of traps exist in terrestrial environments. Perhaps the most remarkable is amber, which is fossilized tree sap. Many specimens of amber contain insects and even small vertebrates upon which sap fell and trapped these animals in the sticky mass. Various sinkholes, such as the Pleistocene Mammoth Site of Hot Springs, South

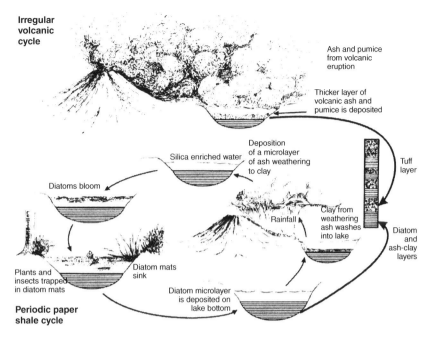

Figure 4.15 Two different cycles of sedimentation were largely involved in preservation of the Eocene Florissant fossils. An irregular volcanic cycle deposited ash and pumice from volcanic eruptions preserving fossil forests on land and depositing thick tuff layers in lakes. In the lakes, deposition typically produced paper shales, consisting of alternating layers of freshwater diatoms that settled to the bottom as part of mucilaginous mats and microlayers of clay formed by the weathering of volcanic clay as it washed into the lake. Plants and insects were trapped in the diatom mats and show exquisite preservation of soft tissues as sediment impressions or as compressions of the original organism, with a darkened color due to the organism's residual carbon content. Many of the compressed fossils retain some of the original three-dimensional nature of the organism. From Meyer (2003). Reproduced with permission from Smithsonian Institution.

Dakota, are examples where animals fell into pits from which they could not escape. Another kind of trap is the petroleum seep, with the most famous being the Pleistocene La Brea Tar Pits, found in Los Angeles (California), which range from 10,000 to 40,000 years old. These deposits exist because petroleum seepage produced shallow pools of tar associated also with ponds of freshwater. Herbivores that came to drink at the ponds inadvertently became stuck in the tar, and their immobilization attracted predators, which also became stuck. Therefore, bones of predators are the most common fossils found in these deposits. These bones were rapidly buried and became impregnated by asphalt,

which aided in their preservation. Plant and insect fossils are also abundant.

Thus, conservation lagerstätten are products of the fortuitous co-occurrence of unique biological, chemical, and sedimentary conditions and are rare exceptions to the usual fate of organic remains as they proceed to possible inclusion in the fossil record. Although additional lagerstätten are being found all the time, lagerstätten with soft tissue preservation are much less common than the more typical fossil deposits. Clearly, the formation of lagerstätten represents some of the more unusual events in the Earth's biological and geological history, and their uniqueness is best exemplified

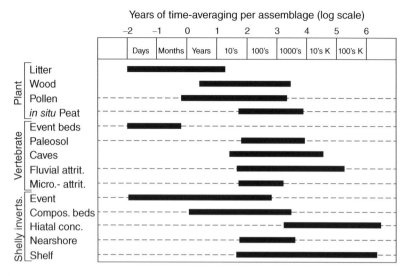

Figure 4.16 Estimated limits on time averaging of selected types of continental plant tissues and vertebrate and marine invertebrate assemblages. The different categories (tissues vs. deposits) reflect the fact that those studying fossil plants regard tissue type as playing the most important role in time averaging for plant remains, while those studying fossil animals regard depositional environments or process as more important. Attritional (attrit.) assemblages involve the accumulation of discarded organic products and input from normal mortality over periods of years to millennia. Microvertebrate fossil assemblages (Micro.) are also called microsites. Composite (compos.) beds reflect more than one sedimentary event during deposition. Hiatal concentrations (conc.) are fossil assemblages deposited during a long-term period of slow sediment accumulation. From Behrensmeyer et al. (2000). Reproduced with permission from Cambridge University Press.

by their scattered occurrence in the sedimentary record of the past 600 million years.

Taphofacies and time averaging

The sedimentary record can be thought of as a recording instrument, which in its own way records not only the history of environments on Earth but also the presence of life in those environments. As we have discussed, some depositional environments produce a coarse representation of that life, while others produce a relatively detailed representation. As defined by Carlton Brett and Gordon Baird, a taphofacies is a body of sedimentary rock with fossils where the fossils have similar taphonomic attributes (Brett and Baird 1986). Thus, the fossils in a taphofacies have experienced the same

taphonomic processes, coming from uniform depositional environments with similar diagenetic processes.

The amounts of time which various fossil assemblages represent can range from days to months to years all the way up to hundreds of thousands of years. As shown in Fig. 4.16 for plant, vertebrate, and shelly invertebrate fossils, which were deposited under different environmental regimes, some of these environments can represent very brief intervals of time, such as litter on the forest floor or vertebrate event beds or shelly invertebrate deposits. But many different sorts of fossil types in different environments can be reworked and only slowly become part of the lithosphere, and Fig. 4.16 shows a broad variety of different types of fossils where assemblages can represent perhaps hundreds of years up to hundreds of thousands of years and even longer.

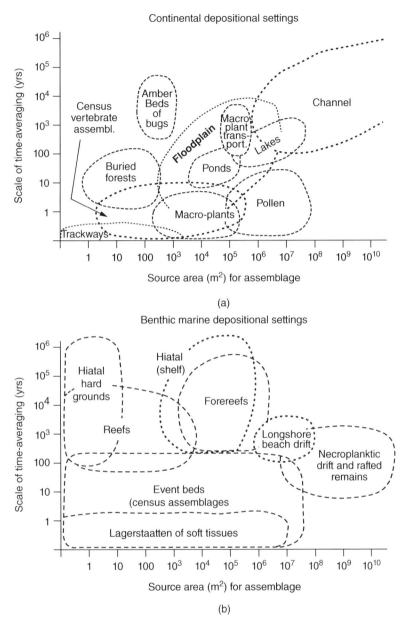

Figure 4.17 Spatial and temporal representation in fossil assemblages for different major groups of organisms in continental and benthic marine depositional settings. (a) Continental settings: dotted lines show areas of the time/space plot occupied by vertebrate remains, and dashed lines plant remains; estimate for pollen excludes trees because certain morphotypes can be transported hundreds of miles by water or thousands of miles by wind prior to settling from the water or air column, respectively. (b) Benthic marine settings include shelly macroinvertebrates and exclude nektonic and planktonic contributions to the fossil assemblage, because spatial resolution of these components can depend upon current drift. From Behrensmeyer et al. (2000). Reproduced with permission from Cambridge University Press.

This concept of time averaging is further illustrated in Fig. 4.17. Here is illustrated how the size of the source area for assemblages can affect the time averaging that one might expect from fossil assemblages. In continental depositional settings, trackways would represent very short intervals of time necessary for their formation. But various other sorts of environmental settings for vertebrates and plants, ranging from buried forests to channel deposits, can represent a broader range of time. For marine deposits, there are various lagerstätten which represent the briefest time intervals for time averaging. Thus, environments where soft tissues are preserved and where the remains could not have been disturbed after they were immediately deposited represent a very short amount of time averaging. But environments such as reefs, which are large constructions and have a lot of recycling of organic remains before final burial into the lithosphere, can represent enormous amounts of time.

Summary

Taphonomic analysis is by necessity a multidisciplinary science. Much remains to be done, but it is remarkable how much has been learned in the past few decades that enables us to understand how fossils under study have been preserved in space and time. One might have once agreed with Darwin that the incompleteness of the fossil record is a strong impediment to understanding the meaning of the fossil record. But we have learned so much on how this incompleteness, basically a taphonomic phenomenon, develops – that our level of certainty on the fundamental biological messages of the fossil record continues to become stronger, a situation that will only keep improving with additional study.

References

Behrensmeyer, A.K. & Kidwell, S.M. 1985. Taphonomy's contributions to paleobiology. *Paleobiology* 11, 105–119.

Behrensmeyer, A.K., Kidwell, S.M. & Gastaldo, R.A. 2000. Taphonomy and paleobiology. *Paleobiology* 26, 103–147.

Brett, C.E. & Baird, G.C. 1986. Comparative taphonomy: A key to paleoenvironmental interpretation based on fossil preservation. *Palaios* 1, 207–227.

Cherns, L., Wheeley, J.R. & Wright, J.P. 2011. Taphonomic bias in shelly faunas through time: Early aragonitic dissolution and its implications for the fossil record. *In* Allison, P.A. & Bottjer, D.J. (eds.), *Taphonomy: Process and Bias Through Time. Topics in Geobiology* 32, Springer, Dordrecht pp. 79–105.

Droser, M.L., Gehling, J.G. & Jensen, S.R. 2006. Assemblage palaeoecology of the Ediacara biota: The unabridged edition? *Palaeogeography, Palaeoclimatology, Palaeoecology* 232, 131–147.

Gastaldo, R.A., Knight, C., Neveling, J. & Tabor, N. 2014. Late Permian paleosols from Wapadsberg Pass, South Africa: Implications for Changhsingian climate. *Geological Society of America Bulletin* 126, 665–679.

Greenwood, D.R. 1991. The taphonomy of plant macrofossils. *In* Donovan, S.K. (ed.), *The Processes of Fossilization*, Columbia University Press, pp. 141–169.

Hendy, A.J.W. 2011. Taphonomic overprints on Phanerozoic trends in biodiversity: Lithification and other secular megabiases. *In* Allison, P.A. & Bottjer, D.J. (eds.), *Taphonomy: Process and Bias Through Time. Topics in Geobiology* 32, Springer, Dordrecht, 79–105.

Martin, R.E. 1999. *Taphonomy: A Process Approach*, Cambridge University Press.

Meyer, H.W. 2003. *The Fossils of Florissant*, Smithsonian Books, Washington, D.C.

Narbonne, G.M., Gehling, J.C. & Vickers-Rich, P. 2007. *The Misty Coasts of Newfoundland. In* Fedonkin, M.A., Gehling, J.G., Grey, K., Narbonne, G.M. & Vickerts-Rich, P. (eds.), *The Rise of Animals: Evolution and Diversification of the Kingdom Animalia*. The Johns Hopkins University Press, pp. 53–67.

Noto, C.R. 2011. Hierarchical control of terrestrial vertebrate taphonomy over space and time: Discussion of mechanisms and implications for vertebrate paleobiology. *In* Allison, P.A. & Bottjer, D.J. (eds.), *Taphonomy: Process and Bias Through Time. Topics in Geobiology* 32, Springer, Dordrecht, 287–336.

Prevec, R., Gastaldo, R.A., Neveling, J., Reid, S.B. & Looy, C.V. 2010. An autochthonous Glossopterid flora with latest Permian palynomorphs and its depositional setting from the *Dicynodon* Assemblage Zone of the southern Karoo Basin, South Africa. *Palaeogeography, Palaeoclimatology, Palaeoecology* 292, 381–408.

Raup, D.M. & Stanley, S.M. 1971. *Principles of Paleontology*. W.H. Freeman and Company.

Tarhan, L.G., Droser, M.L. & Gehling, J.G. 2010. Taphonomic controls on Ediacaran diversity: Uncovering the holdfast origin of morphologically variable enigmatic structures. *Palaios* 25, 823–830.

Zhao, F., Hu, S., Caron, J.-B., Zhu, M., Yin, Z. & Lu, M. 2012. Spatial variation in the diversity and composition of the Lower Cambrian (Series 2, Stage 3) Chengjiang Biota, southwest China. *Palaeogeography, Palaeoclimatology, Palaeoecology* 346–347, 54–65.

Additional reading

Allison, P.A. & Briggs, D.E.G. (eds.). 1991. Taphonomy: Releasing the Data Locked in the Fossil Record. *Topics in Geobiology*, Volume 9. Plenum Publishing Corporation.

Allison, P.A. & Bottjer, D.J. (eds.). 2011. Taphonomy: Process and Bias Through Time. *Topics in Geobiology*, Volume 32. Springer, Dordrecht.

Bottjer, D.J., Etter, W., Hagadorn, J.W. & Tang, C.M. (eds.). 2002. Exceptional Fossil Preservation. Columbia University Press.

Kidwell, S.M. & Behrensmeyer, A.K. 1993. Taphonomic Approaches to Time Resolution in Fossil Assemblages. *Short Courses in Paleontology*, Number 6. The Paleontological Society.

Laflamme, M., Schiffbauer, J.D. & Darroch, S.A.F. (eds.). 2014. Reading and Writing of the Fossil Record: Preservational Pathways to Exceptional Fossilization. *The Paleontological Society Papers*, 20. The Paleontological Society.

5 Bioturbation and trace fossils

Introduction

Bioturbation of sediments is one of the most important sources of paleoecological information from the stratigraphic and fossil record. Organisms utilize a variety of behaviors when they interact with the environment. When those behaviors are played out over or in a substrate, the behavior may result in characteristic structures in the affected sediment, or rock, or organic surface, which are termed trace fossils.

Trace fossils

This preservation of behavior of organisms is a phenomenon not typically accessible from body fossils. Figure 5.1, which outlines invertebrate behavior for the types of organisms that make invertebrate trace fossils, illustrates the wide variety of behaviors that are involved. These different behaviors such as escape, resting, or predation have been given names for the trace fossils that they produce. For example, fugichnia are escape burrows, agrichnia are farming burrows, and domichnia are dwelling burrows. These different types of behaviors that produce different kinds of trace fossils have also led to scientists giving many of them names using Linnaean nomenclature for the trace fossils. Thus, in Fig. 5.1, the branching domichnia example is also known using the taxonomic system for trace fossils as the ichnogenus *Ophiomorpha*. This approach of classifying trace fossils using Linnaean nomenclature, rather than just assigning them to the organisms that (may have) created them, is typically employed because the makers of a number of trace fossils are not fully understood and many organisms exhibit a variety of behaviors, thus making more than one type of trace fossil. The trace fossil record of these different types of behaviors can commonly be found in marine and terrestrial sedimentary rocks and provides a vast source of information on how animals lived and interacted with each other and the physical and chemical environments surrounding them.

Marine environments

As is the case for body fossils, the processes whereby biogenic sedimentary structures become part of the stratigraphic record are complex. As an example, Fig. 5.2 shows the various processes which trace fossils and other biogenic sedimentary structures in marine pelagic/hemipelagic muds pass through as they become a permanent part of the lithosphere. The mixed layer has the highest water content and is commonly burrowed and reburrowed by mobile as well as sedentary organisms. Thus, in the mixed

Paleoecology: Past, Present and Future, First Edition. David J. Bottjer.
© 2016 John Wiley & Sons, Ltd. Published 2016 by John Wiley & Sons, Ltd.

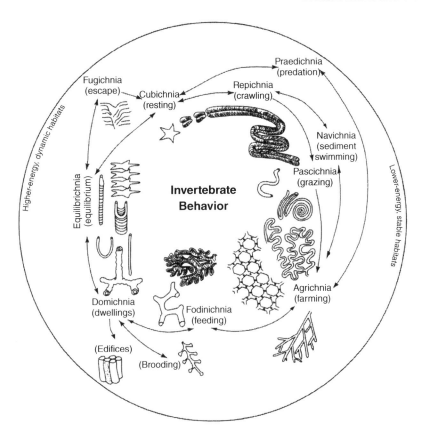

Figure 5.1 Trace fossils made by invertebrates represent a suite of behaviors that may combine or grade with each other. In this schematic, the ethological patterns reflected as typical trace fossils are grouped so that the families of feeding behaviors are evident. Behaviors and typical trace fossils associated with higher-energy environmental settings are depicted on the left, while those typical of lower-energy environments are found on the right. This association of behaviors and resulting trace fossils with depositional processes forms the basis for the ichnofacies classification. From MacEachern et al. (2010). Reproduced with permission from the Geological Association of Canada.

layer, discrete trace fossils with sharp outlines are not commonly preserved, but more commonly, the record of bioturbation is shown as a mixing of sediment which is typically preserved as sedimentary fabric, which will be discussed later. The underlying transition layer can be a number of centimeters below the sediment–water interface, and this layer has much less water content, as it represents the beginning of compaction so that individual burrows can be preserved as discrete biogenic sedimentary structures or trace fossils. The transition layer can be subject to additional bioturbation. Below that is

the historical layer where trace fossils are preserved as they would appear ultimately in the stratigraphic record. This layer is also where concretions can grow and diagenesis, along with increased compaction and reduction in water content, continues to occur.

Dolf Seilacher recognized that particular behaviors, as represented by discrete trace fossils, are more common in certain environments than in others. This observation led to the concept of ichnofacies, similar to the facies concept in sedimentology, where the suites of trace fossils found in sedimentary rocks were controlled by the environmental conditions

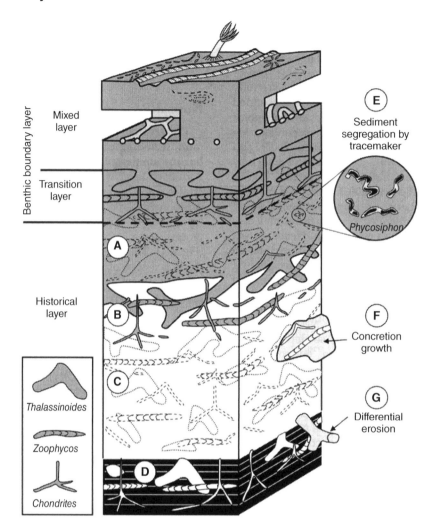

Figure 5.2 Sediment accumulation and preservation of trace fossils in marine pelagic/hemipelagic mud. Under typical depositional conditions, mixed layer structures are not preserved. Structures formed in the transition layer dominate the historical layer, although visibility of these structures can be limited by poor contrast between trace fossils and host sediments (A, C). In contrast, these transition-layer structures may have enhanced visibility due to a variety of processes including changes in sediment type and downward transport of contrasting sediments, such as at bed junctions (B, D); textural or compositional segregation by tracemakers (E); diagenetic mineralization (F); and differential weathering/erosion (G) of burrows or host sediments. Trace fossils preserved at bed-junction preservation show particularly good preservation if they crosscut previously unbioturbated sediments such as dark, laminated shales (D). From Savrda (2007). Reproduced with permission from Elsevier.

under which they were deposited and thus can be helpful in reconstructing ancient environmental settings. The classical ichnofacies first promulgated by Seilacher ranged from nearshore marine environments to the shelf to bathyal and deep sea (Fig. 5.3). These various ichnofacies can be of great utility when one is studying the depositional environments of sedimentary rocks. Of course, they

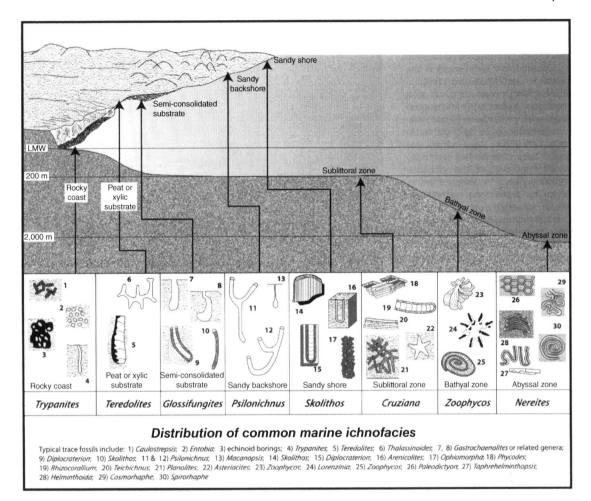

Figure 5.3 Schematic of common marine ichnofacies, which vary from shallow to deepwater environments and substrate type. Sketches of typical trace fossils for each ichnofacies are shown; the name of each ichnofacies (e.g., *Trypanites*) reflects a characteristic trace fossil for that ichnofacies. To identify a particular ichnofacies, the trace fossil which gives that ichnofacies its name does not need to be present. From MacEachern et al. (2010), where data sources are indicated. Reproduced with permission from the Geological Association of Canada. (*See insert for color representation.*)

do not stand alone as indicators of sedimentary environment, but together with other physical and chemical sources of data are extremely valuable in understanding where sediments in the sea as well as on land have been deposited.

Not all biogenic sedimentary structures are discrete identifiable trace fossils. Much of what we see in the biogenic structure realm are indistinct indicators that sediments have been bioturbated in the mixed layer. To include distinct and indistinct biogenic structures, the overall record of bioturbation in sedimentary rocks has been termed ichnofabric. Thus, ichnofabric includes identifiable trace fossils as well as the bioturbated texture commonly produced through mixing of sediments.

Different types of behaviors and thus trace fossils can produce ichnofabrics with different textures. Some sedimentary rocks are thoroughly bioturbated, while some have just a burrow or two as their record of bioturbation. Approaches to measure the extent of bioturbation recorded in a vertical sedimentary sequence have been varied. Several schemes have been proposed to assess the amount of bioturbation recorded in sedimentary rock as reflected by ichnofabric.

Figure 5.4 shows examples of the ichnofabric index method for several different settings, including shelf and shallower nearshore environments characterized by the *Skolithos* and *Ophiomorpha*

ichnofabrics, as well as deep-sea ichnofabric. In work done with Mary Droser, these various ichnofabric indices were defined as a system of flashcards where original primary physical sedimentary structures without bioturbation are indicated by ichnofabric index 1 and then increasing destruction of these physical structures by bioturbation is represented by an ichnofabric index of 2–5, with complete homogenization by bioturbation being 6. Figure 5.5 shows an example of ichnofabric index 5 from an Upper Cretaceous chalk. Various applications

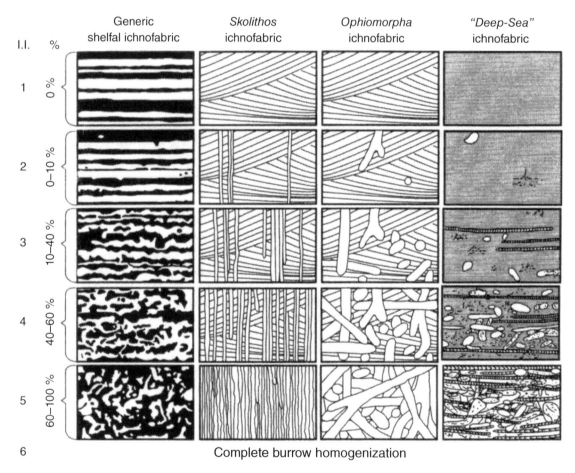

Figure 5.4 Ichnofabric indices exemplified by flashcards according to the method proposed by Droser and Bottjer (1989). These show the amount of sediment reworked for different marine sedimentary environments as observed in vertical cross section. Percent of reworking for ichnofabric indices (I.I.) 1–5 shown on left. Ichnofabric index 1 is a fabric that is all physical sedimentary structures (lacking bioturbation), and ichnofabric index 6 is complete homogenization by bioturbation. *Skolithos* and *Ophiomorpha* ichnofabrics are for sandy nearshore sedimentary environments. From McIlroy (2004). Reproduced with permission from the Geological Society.

Figure 5.5 Thorough bioturbation resulting in ichnofabric index (ii) 5 developed at a marly chalk–chalk transition (mid- to outer shelf) in the Upper Cretaceous (Campanian) Demopolis Chalk (Selma Group) exposed in western Alabama (eastern Gulf coastal plain), United States. Distinct lighter-filled trace fossils include *Thalassinoides* (wide vertical structure on left side and probably highest long horizontal structure near top), *Zoophycos* (horizontal structure with faint meniscate backfill below left side of horizontal *Thalassinoides*?), and *Chondrites* (lightest structures including dots to short branching segments). Additional information can be found in Locklair and Savrda (1998). Scale is in centimeters. Photograph by Charles E. Savrda. Reproduced with permission. (*See insert for color representation.*)

of ichnofabric indices have been made where the goal was to understand patterns in the amount of bioturbation recorded in different packages of sedimentary rocks.

Figure 5.6 shows a schematic column of sedimentary rocks representing shelf environments where 5 m of rock have been logged through the column for the presence of different ichnofabric indices. These ichnofabric index data are then summed by producing a histogram of the percent of the studied stratigraphic interval which contains different ichnofabric indices. This histogram is termed an ichnogram, and it can give a quick understanding of the amount of bioturbation which that particular stratigraphic interval has undergone.

Although ichnofabric indices are very useful for understanding sedimentary rocks in vertical sequence analysis, commonly bioturbation structures are exposed on bedding planes where not only might there be well-developed trace fossils but also where the extent of bioturbation on the bedding plane varies. To assess this amount of bioturbation on bedding planes, a bedding plane bioturbation index has also been developed along with flashcards as illustrated in Fig. 5.7. The bedding plane bioturbation index and other approaches that can be more quantitative are very valuable in determining trends in bioturbation and have been used, for example, in understanding the development and evolution of the early history of marine bioturbation in the Cambrian (Marenco and Bottjer 2011).

Of special interest to understanding evolutionary paleoecology is the record of predation traces made by durophagous predators. These include boreholes made by various invertebrate predators such as

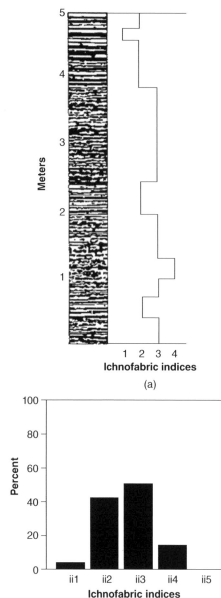

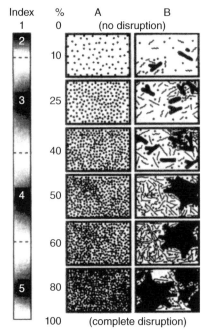

Figure 5.7 The bedding plane bioturbation index, with different indices showing the proportion of bedding planes covered by trace fossils. Column A represents example bedding planes covered by trace fossils of even size and shape with even distributions. Column B represents example bedding planes covered by trace fossils of different sizes and shapes and with uneven distributions. From McIlroy (2004). Reproduced with permission from the Geological Society.

Figure 5.6 Measurement of ichnofabric indices (ii). (a) Schematic diagram of hypothetical stratigraphic section logged for ichnofabric indices during vertical sequence analysis. (b) Ichnogram computed from this hypothetical data. Logging is based on a 50 cm-wide field of view. Of 5 m measured, ii1 was recorded from 15 cm, ii2 from 200 cm, ii3 from 255 cm, and ii4 from 30 cm. From Droser and Bottjer (1993).

gastropods and cephalopods and bite marks on various shells and bones. Studies of the occurrence of predation traces through the Phanerozoic have been particularly valuable toward understanding the early evolution of predation in the Late Ediacaran and subsequent predation trends in the Phanerozoic, such as those characteristic of the Mesozoic Marine Revolution (e.g., Chapter 12; Fig. 12.13).

Other significant processes leading to destruction of hard substrates in marine environments include extensive production of macroscopic borings by such organisms as sponges, polychaetes, bivalves, and barnacles. Another component of bioerosion includes microborings made by cyanobacteria, algae, and fungi which occur extensively and are a significant mechanism in the destruction of carbonate skeletons. Wood can be found in nearshore environments and also as driftwood, and such woody substrates are typically bored by bivalves in the family Teredinidae, commonly known as the shipworms.

Terrestrial environments

Bioturbation has also been an active process in terrestrial environments since animals colonized land by the Ordovician and various other nonmarine settings later in the Paleozoic. The study of terrestrial bioturbation has lagged that of marine bioturbation, but much has been learned over the past several decades. The ichnofacies concept has been expanded and refined toward terrestrial applications, as shown in Fig. 5.8. These nonmarine ichnofacies, largely based on the distribution of trace fossils made by invertebrates, can also be very

useful in determining depositional environments of sedimentary rocks.

One of the unique characteristics of bioturbation in terrestrial environments is that there is a large record of vertebrate tracks and trackways. This information not only records the presence of vertebrates in particular environments but also provides behavioral information. Of particular importance is the dinosaur track record. Figure 5.9 shows various trackways of saurischians and the characteristics which they can display. Figure 5.10 illustrates the kinds of trackways that are valuable for understanding ornithischian dinosaurs. Much information on behavior of dinosaurs has been gained, such as evidence on social behavior (Figs. 5.11 and 11.10) and predator–prey interactions as well as the gaits and speed of various dinosaur groups. Significant information is also being gained from study of other Mesozoic vertebrate trackways, such as those of pterosaurs, where the quadrupedal behavior of many of these flying reptiles is demonstrated through their

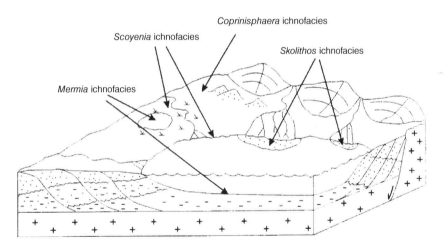

Figure 5.8 Ichnofacies distribution in terrestrial environments, including lacustrine and fluvial settings. These are largely based on the occurrence of trace fossils made by invertebrates. The *Mermia* ichnofacies characterizes subaqueous oxygenated lacustrine environments and includes grazing and feeding traces as well as locomotion traces. The *Scoyenia* ichnofacies is found in environments that are periodically covered in water, such as the margins of fluvial and lacustrine systems. This ichnofacies is characterized by traces of mobile organisms including tracks, trails, and meniscate-backfilled burrows. The *Skolithos* ichnofacies, with simple vertical as well as U-shaped burrows, is found in high-energy lacustrine and fluvial environments. The *Coprinisphaera* ichnofacies is present in paleosols. From Buatois and Mangano (2004). Reproduced with permission from the Geological Society.

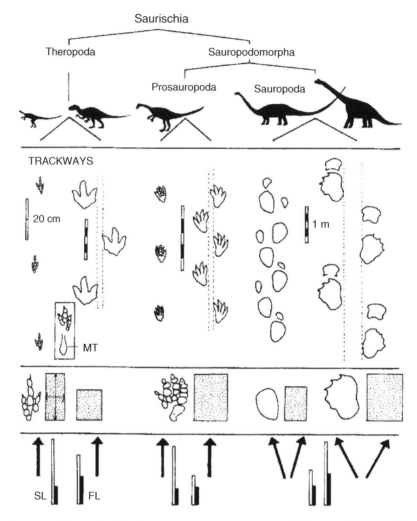

Figure 5.9 Trackways made by saurischian dinosaurs. The Saurischia, one of the two major groups of dinosaurs, comprises the herbivorous sauropodomorphs, which include the sauropods such as *Diplodocus* and their ancestral relatives, as well as the carnivorous theropods such as *Tyrannosaurus*. Trackway configurations in the center for these main groups of saurischian dinosaurs (top), with an inset showing a theropod footprint with metatarsal impressions (MT). From left to right, trackways are *Grallator*, *Megalosauripus*, *Otozoum* narrow-gauge, *Otozoum* wide-gauge, *Parabrontopodus*, and *Brontopodus*, all based on type material. All scale bars 1 m except for *Grallator*. Note that all footprints are longer than wide and that footprint axes (arrows) point forward or outward (bottom rows). Characteristic step and foot length (SL and FL) ratios are also shown, and stippled rectangles correspond to track length and width. From Lockley (2007). Reproduced with permission from the SEPM Society for Sedimentary Geology.

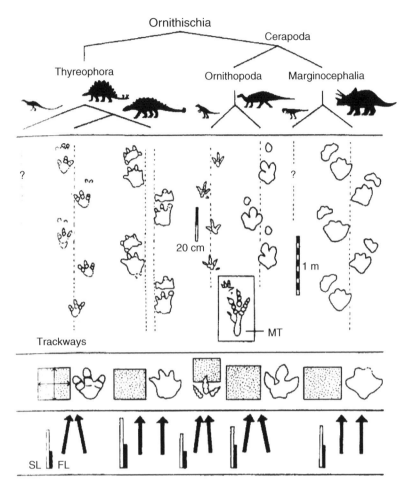

Figure 5.10 Trackways made by ornithischian dinosaurs. The Ornithischia, one of the two major groups of dinosaurs, are all herbivores and include such familiar animals as *Stegosaurus*, *Triceratops*, and *Edmontosaurus*. Trackway configurations in the center for main groups of ornithischian dinosaurs (top), with an inset showing an ornithopod footprint with metatarsal impressions (MT). From left to right, trackways are unnamed stegosaur trackway, *Tetrapodosaurus*, *Anomoepus*, *Caririchnium*, and *Ceratopsipes*, all based on type material. All at same scale except for *Anomoepus*. ? = no trackways known for scutellosaurids and pachycephalosaurids. Note that all footprints are as wide as or wider than long and that footprint axes (arrows) point inward and forward (bottom rows). Characteristic step and foot length (SL and FL) ratios are also shown, and stippled rectangles correspond to track length and width. From Lockley (2007). Reproduced with permission from the SEPM Society for Sedimentary Geology.

Figure 5.11 Two parallel sauropod trackways from the Upper Cretaceous of Bolivia. The direction of progression is down dip, toward the bottom of the photograph. There also is a theropod trackway oriented from right to left and crossing the lower third of the sauropod trackways. These are from the Cal Orcko quarry site on the outskirts of Sucre, Bolivia. They occur on a limestone bedding plane that is part of a mosaic of lacustrine and marginal lacustrine facies of the Upper Cretaceous El Molino Formation (Maastrichtian). Nearby sites in the El Molino Formation have associated sauropod trackways indicating that groups of at least 11 individuals were moving together. Collectively, this evidence suggests that sometimes these animals traveled in social groups. Additional information can be found in Lockley et al. (2002). Photograph by Martin Lockley. Reproduced with permission. (*See insert for color representation.*)

unique tracks. The same can be said for Mesozoic and Cenozoic bird tracks, as well as extensive Cenozoic mammal trackways that reveal behavioral traits of large mammals such as mastodons, mammoths, camels, horses, and humans. Other sedimentary structures associated with vertebrate behavior include coprolites and nests like those made by dinosaurs (e.g., Figs. 11.8 and 11.9).

Summary

The record of bioturbation provides an enormous addition to the record of body fossils when one is attempting to understand the paleoecology of an ancient sedimentary environment. It contains much of the record of soft-bodied invertebrates that existed in an environment and how they interacted ecologically, particularly for the infauna. For animals that walk on and deform the sediment surface, this record is essential for understanding much about behavior that cannot be obtained in any other way. Therefore, paleoecological studies that combine analyses of body fossils and bioturbation structures can be particularly powerful.

References

Buatois, L.A. & Mangano, M.G. 2004. Animal-substrate interactions in freshwater environments: Applications of ichnology in facies and sequence stratigraphic analysis of fluvio-lacustrine successions. *In* McIlroy, D. (ed.), *The Application of Ichnology to Palaeoenvironmental and Stratigraphic Analysis. Geological Society Special Publication* **228**, 311–333.

Droser, M.L. & Bottjer, D.J. 1989. A semiquantitative field classification of ichnofabric. *Journal of Sedimentary Petrology* 56, 558–559.

Droser, M.L. & Bottjer, D.J. 1993. Trends and patterns of Phanerozoic ichnofabrics. *Annual Reviews of Earth and Planetary Sciences* 21, 205–225.

Locklair, R.E. & Savrda, C.E. 1998. Ichnology of rhythmically bedded Demopolis Chalk (Upper Cretaceous, Alabama): Implications for paleoenvironment, depositional cycle origins, and tracemaker behavior. *Palaios* 13, 423–438.

Lockley, M.G. 2007. The morphodynamics of dinosaurs, other archosaurs, and their trackways: Holistic insights into relationships between feet, limbs, and the whole body. *In* Bromley, R.G., Byatois, L.A., Mangano, G., Genise, J.F. & Melchor, R.N. (eds.), *Sediment-Organism Interactions: A Multifaceted Ichnology. SEPM (Society for Sedimentary Geology) Special Publication* **88**, 27–51.

Lockley, M., Schulp, A.S., Meyer, C.A., Leonardi, G. & Kerumba Mamani, D. 2002. Titanosaurid trackways from the Upper Cretaceous of Bolivia: Evidence for large manus, wide-gauge locomotion and gregarious behaviour. *Cretaceous Research* 23, 383–400.

MacEachern, J.A., Pemberton, S.G., Gingras, M.K. & Bann, K.L. 2010. Ichnology and facies models. *In* James, N.P. & Dalrymple, R.W. (eds.), *Facies Models 4.* Geological Association of Canada, 19–58.

Marenco, K.N. & Bottjer, D.J. 2011. Quantifying bioturbation in Ediacaran and Cambrian Rocks. *In* Laflamme, M., Schiffbauer, J.D. & Dornbos, S.Q. (eds.), *Quantifying the Evolution of Early Life: Numerical Approaches to the Evaluation of Fossils and Ancient Ecosystems. Topics in Geobiology* 36, 135–160. Springer, Dordrecht.

McIlroy, D. (ed.). 2004. *The Application of Ichnology to Palaeoenvironmental and Stratigraphic Analysis, Geological Society Special Publication No. 228.* The Geological Society, London, 3–27.

Savrda, C.E. 2007. Taphonomy of trace fossils. *In* Miller, W.M., III, (ed.), *Trace Fossils: Concepts, Problems, Prospects*, Elsevier, Amsterdam, 92–109.

Additional reading

Buatois, L.A. & Mangano, M.G. 2011. *Ichnology: Organism–Substrate Interactions in Space and Time.* Cambridge University Press.

Hasiotis, S.T. 2002. *Continental Trace Fossils, SEPM Short Course Notes No. 51.* SEPM (Society for Sedimentary Geology).

Lockley, M. & Hunt, A.P. 1995. *Dinosaur Tracks and Other Fossil Footprints of the Western United States.* Columbia University Press.

Martin, A.J. 2014. *Dinosaurs Without Bones: Dinosaur Lives Revealed by their Trace Fossils.* Pegasus

Miller, W., III. (ed.). 2007. *Trace Fossils: Concepts, Problems, Prospects.* Elsevier.

Seilacher, A. 2007. *Trace Fossil Analysis.* Springer.

6 Microbial structures

Introduction

The presence, importance, and activity of microbes in ancient ecosystems can commonly be understood through analysis of the sedimentary structures which they induce. Some of these structures have been considered to be trace fossils, but their importance and mode of construction are such that herein they are treated separately. As we have learned more about these structures, they have become a significant source of information on the microbial component of many ecosystems, and they have come to be known collectively as microbialites.

Biofilms

Microbes form biofilms on living and nonliving surfaces in most environments on Earth. A significant component of these biofilms is extracellular polymeric substance (EPS), which is produced by and surrounds the microbes. EPS consists of polysaccharides and proteins that protect the microbes living within it. In sedimentary environments, on land and in the sea, these biofilms form conditions that produce structures which are preserved in the sedimentary record, and, as has already been discussed, can play a major role in taphonomic processes.

In mineral-precipitating marine environments, biofilms that develop into microbial mats can form laminated structures termed stromatolites (Fig. 6.1). Biofilms also produce related carbonate microbialites such as thrombolites, which have a clotted rather than laminated internal structure. In siliciclastic sediments, these structures do not grow by mineral precipitation, so they are typically more diminutive, forming "microbially induced sedimentary structures" (MISS; Fig. 6.1), including such bedding plane features as wrinkle structures or elephant skin. Some of these siliciclastic structures have also been termed "textured organic surfaces" (Gehling and Droser 2009). In arid terrestrial environments, bacteria as well as fungi and algae can form a partially cemented surface called cryptobiotic crust or soil. Production of microbial structures is favored where there is minimal sediment disturbance, so wherever there is significant bioturbation, on land or in the sea, the potential to form microbialites decreases.

Carbonate environments

Stromatolites are common features of Precambrian and early Paleozoic marine carbonate rocks, before the evolution of significant bioturbation. Stromatolites differ from laminated sediments because they

Paleoecology: Past, Present and Future, First Edition. David J. Bottjer.
© 2016 John Wiley & Sons, Ltd. Published 2016 by John Wiley & Sons, Ltd.

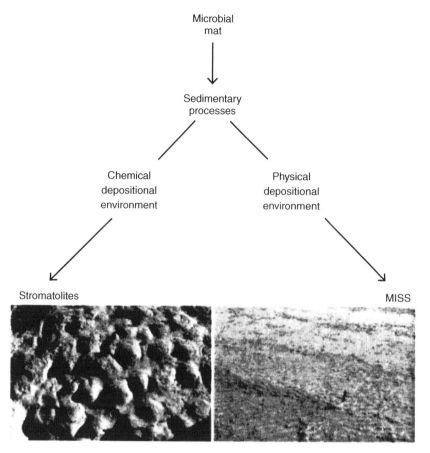

Figure 6.1 Pathways toward deposition of microbial structures in sedimentary rocks. Microbial mats, which are constructed by benthic cyanobacteria and other microorganisms, serve as the biotic template for these structures. The sedimentary processes that interact with the microbial mat can then follow one of two pathways. In carbonate ("chemical") depositional environments, microbial mats induce the formation of stromatolites, which includes the precipitation of calcium carbonate and the subsequent development of positive topography on the seafloor. In siliciclastic ("physical") environments, where mineral precipitation or cementation play no role, microbial mats form "microbially induced sedimentary structures" (MISS), which typically appear in sedimentary outcrops as features that are more subtly detected than are stromatolites. Wrinkle structures are a common MISS, as are multidirected ripple marks. The photo on the left shows typical, domal stromatolites (scale: 10 cm). The photo on the right shows multidirected ripple marks (scale knife 25 cm). From Noffke (2010). Reproduced with permission from Springer Science and Business Media.

form topographically positive surfaces (Fig. 6.2) that grow in a variety of shapes ranging from mounds to branches.

In sedimentary sequences, stromatolites are typically seen as laminated structures with positive topography that at times can range up to meters in height. Of course, many structures that would be called stromatolites may be precipitated inorganically in sedimentary and also diagenetic environments. But stromatolites that are influenced by microbial mat presence are common and can be studied in a variety of ways.

Figure 6.2 A group of columnar stromatolites that are part of a bioherm preserved in dolostone. Individual stromatolites are on the order of tens of centimeters tall, and the bioherm extends laterally for several meters. From the Ediacaran Deep Spring Formation at Mt Dunfee, Nevada, United States. Additional information can be found in Oliver and Rowland (2002). Photograph by Russell S. Shapiro. Reproduced with permission. (*See insert for color representation.*)

Figure 6.3 shows an outline of the broad mechanisms involved in producing modern stromatolites and how they can be studied. Examining them at the microscale enables a determination of characteristics of different microbial communities that lead to the development of stromatolites. The mesoscale is the production of visible sedimentary laminations characteristic of stromatolites. These laminations can be arranged into a variety of shapes and morphologies, and it is largely the broad physical environmental influences under which stromatolites grow that control overall stromatolite morphology. Macrobiological phenomena such as the presence of algae or invertebrates such as macroborers might be present along with the microbial mats. These conditions can also influence the development of morphology and degrees of linkage between stromatolites.

Adding analyses of how ancient stromatolites formed allows a more detailed understanding of the processes which have produced stromatolites and their laminations (Fig. 6.4). EPS is sticky and acts in concert with microbial filaments to trap sedimentary particles from the water column, which help to build the positive aspects of the topography of stromatolites. The stickiness of the microbial mat can trap sedimentary grains transported across the surface of the mat, and subsequently, these grains are bound into the structure by carbonate cementation. Because stromatolites are formed in chemical environments where carbonates can be precipitated, calcium carbonate cements can also be a significant component of stromatolites. Additional stromatolite growth mechanisms include lateral growth of microbial mats over sediment that has accumulated in underlying

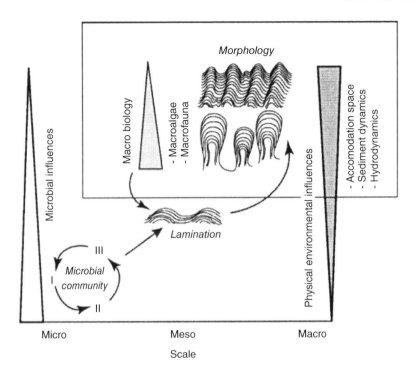

Figure 6.3 Conceptual model for stromatolite formation at different scales based on studies from modern environments. The microscale is dominated by the microbial community which produces mesoscale laminations, and the macroscale includes variations in the shape or morphology of the laminated stromatolite. Different microbial mat communities (I, II, III) are involved in trapping and binding sediment transported along the seafloor as well as precipitating carbonate and carbonate cements. Microbial influence decreases toward the macroscale, and at this level, physical environmental factors increasingly control the formation of stromatolites, including accommodation space, or the space available for sediment deposition, which is controlled by water depth. Other contributing environmental factors influencing stromatolite shape and morphology are sediment dynamics and hydrodynamics. Macrobiology, in particular colonizing macroalgae and boring macrofauna, acts upon the stromatolite ecosystem at times to strongly influence shape and morphology. From Andres and Reid (2006). Reproduced with permission from Elsevier.

depressions and through growth of microbial body fossils (Fig. 6.4).

Thus, stromatolites are complex bioconstructions that range in size from millimeters to centimeters to meters with shapes that are domal to columnar (Fig. 6.2) to conical (Fig. 6.5) to branching. Further study of ancient stromatolites has revealed an even more complex suite of morphologies than can be portrayed in general models for their formation. Quite naturally, these interesting stromatolite morphologies have been studied intensively apart from the general example of stromatolite development. One example studied by Frank Corsetti and John Grotzinger is what has come to be known as tubular or tube-forming stromatolites, in which the laminated stromatolite intervals are so closely adjacent to each other that the areas in between produced tubelike structures, though these are not constructed by stromatolitic processes (Figs. 6.6 and 6.7). To produce these tube-forming stromatolites, the former presence of a somewhat tufted microbial mat is proposed. In this way, sediment would

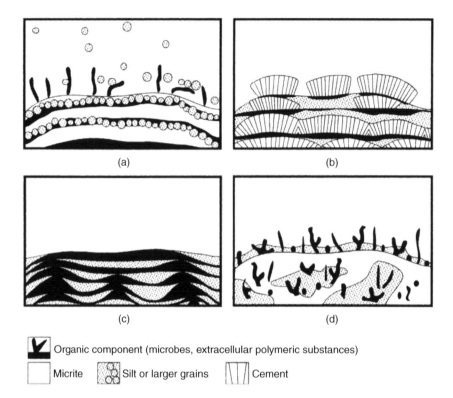

Figure 6.4 Models for stromatolite accretion based on studies of modern and ancient settings. (a) Successive generations of extracellular polymeric substance and filamentous microbes forming a microbial mat create a sticky surface that leads to trapping and binding of grains of carbonate mud, silt, and sand. (b) Cements can be precipitated either directly upon decaying organic compounds or adjacent to the organics with little or no clastic sedimentary component. (c) Successive layers of mechanical sediment can accumulate in depressions with subsequent spread by microbial communities over these sediments, sealing the deposit. (d) Microbial body fossils (external or internal molds, casts, permineralized sheaths) can also be a large component of stromatolites. From Shapiro (2007). Reproduced with permission from Elsevier.

fill the low areas between the tufts, creating the base of a tube. This structure would then grow vertically leading to development of a tube-forming stromatolite.

Siliciclastic environments

Leaving carbonate environments and moving to siliciclastic settings, this is a suite of environments in which until relatively recently there was only limited understanding of how the presence of microbial mats could be detected in ancient sedimentary environments. But work with James Hagadorn on ancient wrinkle structures (Fig. 6.8) as well as work by others on modern and ancient environments has led to the recognition of a variety of bedding plane structures which Nora Noffke has termed MISS. Microbes are always present in surface sediments, and the development of a surficial microbial mat can lead to trapping and binding of distinctive sediments such as heavy minerals. Microbial mats can also lead to changes in rheological properties that affect how surface sediment is formed into

Figure 6.5 Conical stromatolites from the Belt Supergroup (ca. 1.4 billion years ago) in southern Alberta, Canada. Conical stromatolites have typically been considered as strong evidence for biogenic formation, as such shapes have not been modeled without the activity of life. Field of view is 30 cm wide. Photograph by Frank Corsetti. Reproduced with permission. (*See insert for color representation.*)

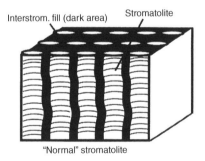
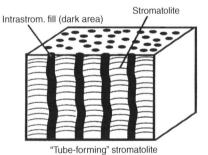

Figure 6.6 Schematic representation showing formation of tube-forming stromatolites. Laminated areas represent stromatolitic lamination, and dark areas represent sedimentation that occurred adjacent to the stromatolites. More widely spaced stromatolites (left) lead to development of stromatolites. Closely spaced stromatolite growth (right) produces tubes of intrastromatolite sediment fill. Note that the tube-forming stromatolite and the more normal stromatolite appear similar in vertical cross section but different in horizontal cross section. In plan view, the tube-forming stromatolite consists of an interconnected network of microbialite with ovate intrastromatolite fill (the tubes; see Fig. 6.7), while the normal stromatolite consists of isolated microbialite separated by interstromatolite fill. From Corsetti and Grotzinger (2005). Reproduced with permission from the SEPM Society for Sedimentary Geology.

Figure 6.7 Bedding plane view of typical Neoproterozoic tube-forming stromatolites from the Noonday Dolomite at Winters Pass Hills, in the Death Valley area, eastern California, United States. Tubes are filled with dark micrite/microspar. The Noonday Dolomite is a so-called cap carbonate deposited in the aftermath of a "Snowball Earth" glacial episode and rests sharply on underlying glacial strata. Additional information can be found in Corsetti and Grotzinger (2005). Photograph by Frank Corsetti. Reproduced with permission. (*See insert for color representation.*)

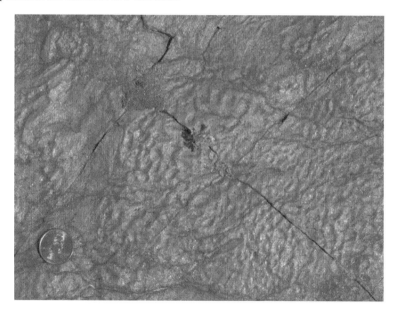

Figure 6.8 Bedding plane surface of the lower Cambrian Harkless Formation with quasi-polygonal, moderately flat-topped wrinkle structures. Associated sedimentary structures indicate deposition in shallow subtidal conditions. Locality is near Cedar Flats, White-Inyo Mountains, CA, United States. Additional information can be found in Hagadorn and Bottjer (1997, 1999) and Bailey et al. (2006). Photograph by Stephen Q. Dornbos. Reproduced with permission. (*See insert for color representation.*)

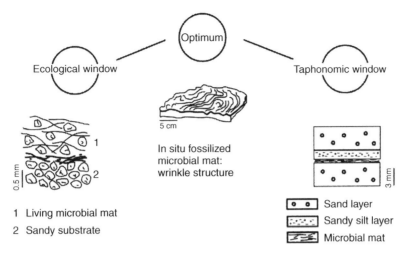

Figure 6.9 Optimum conditions for formation and preservation of microbially induced sedimentary structures (MISS) such as wrinkle structures in siliciclastic environments. These include an ecological window which consists of a sandy substrate that favors the development of thick microbial mats and a taphonomic window which leads to preservation of features of the mats. For the ecological window, microbial mats are best formed in sediments composed of "clear" (translucent) quartz minerals of fine sand grain sizes because these quartz grains allow light to penetrate deeper into the sediment allowing cyanobacteria in the mat to live at greater depths, thus producing a thicker mat. For the taphonomic window, MISS such as wrinkle structures are best preserved if a layer of fine sand and silt is deposited on the established microbial mat, which has accumulated through baffling, trapping, and binding finer-grained sediment particles. Because the mat biostabilizes its substrate, it is not destroyed when it is buried, thus allowing this surface with its mat structures to be preserved. Further taphonomic features such as compaction can modify the surface expression of MISS such as wrinkle structures, so that the wrinkled appearance can represent a combination of primary and secondary processes. From Noffke (2009). Reproduced with permission from Elsevier.

sedimentary structures or broken by erosive processes or subaerial drying. Microbial mats through their development can also produce a vertical topography of tufts and ridges. Interactions of these processes are thought to commonly preserve in the rock record what are termed wrinkle structures (Fig. 6.8). Figure 6.9 illustrates the optimum conditions hypothesized for formation and preservation of MISS such as wrinkle structures in shallow photic zone environments. These structures typically form in sandy substrates and are thought to be best preserved after the microbial mat is initially overlain by a sandy silty event layer. In addition, experimental work has shown that the movement by small amplitude waves of fragments of microbial mats can form wrinkle structures on the seafloor surface (Mariotti et al. 2014).

A variety of MISS and the processes which are proposed to lead to their development and preservation are shown in Fig. 6.10. Biostabilization can lead to stiffening or hardening of surface sediment so that under erosive currents, this surface layer peels off into roll-ups or breaks into mat chips. Gas production due to degradation of buried mats can cause development of dome structures or sponge pore fabrics in such biostabilized sediments. Dewatering of these sediments can also produce distinctive cracking patterns or petee structures. Growth of microbial mats in siliciclastic settings is a primary factor in deposition of laminations, particularly in

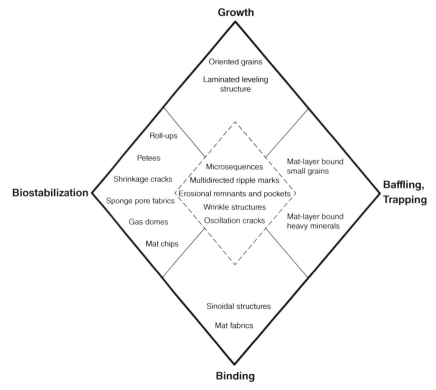

Figure 6.10 Processes which lead to the formation of MISS, including growth, biostabilization, baffling and trapping, and binding, and representative MISS caused by these processes. The center of the diagram includes MISS caused by the interference of all of these processes. From Noffke (2010). Reproduced with permission from Springer Science and Business Media.

deeper environments. Baffling and trapping can produce distinctive sedimentary layers which can be discerned microscopically, and binding produces characteristic structures and fabrics (Fig. 6.10). Basically, many of the same sorts of processes involved in developing stromatolites, except for the precipitation of calcium carbonate minerals, are proposed for the formation of MISS. Of course, this makes MISS structures more diminutive than many stromatolites, but they are commonly easily identifiable on outcrop, particularly bedding planes.

In areas where dewatering and erosional currents are active, a broad variety of MISS is possible. This is typical in modern siliciclastic tidal flat areas where disturbance by bioturbation is commonly reduced due to the harsh conditions. Figure 6.11 shows a proliferation of sedimentary structures that are included as MISS and which are interpreted as indicative of the presence of microbial mats in the original intertidal to shallow subtidal environments. These modern lower intertidal and subtidal environments show the occurrence of wrinkles, and this is one of the most common MISS to be found in ancient subaqueous settings.

Wrinkle structures can have various dimensions and surface patterns of various shapes and sizes and are indicative of the former presence of a microbial mat. Wrinkle structure morphology also reflects the effects of compaction of overlying sediments on the wrinkled surface and the distortion that produces. Because they lack bioturbation and the disturbance which that causes, deeper-water Precambrian

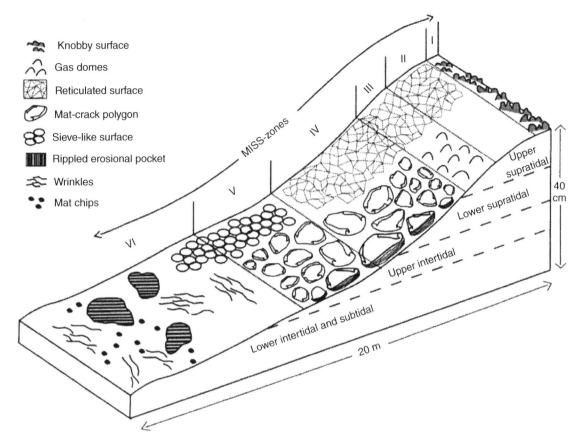

Knobby surface

Gas domes

Reticulated surface

Mat-crack polygon

Sieve-like surface

Rippled erosional pocket

Wrinkles

Mat chips

MISS-zones

I
II
III
IV
V
VI

Upper supratidal

Lower supratidal

Upper intertidal

Lower intertidal and subtidal

40 cm

20 m

Figure 6.11 Modern MISS from shoreline environments in Texas (United States). The supratidal area to upper subtidal area is divided into six MISS zones. Zones I and II contain only knobby surfaces and reticulated surfaces, respectively. Along with reticulated surfaces, zones III and IV contain gas domes and mat cracks, respectively. Zone V is characterized by sieve-like surfaces and mat cracks, whereas zone VI is defined by sieve-like surface cracks and abundant structures such as wrinkles, erosional pockets, and mat chips. Occurrences of these MISS zones are related to the tidal zonation, but the boundaries between the MISS zones may not coincide with that of the tidal zonations. From Bose and Chafetz (2009). Reproduced with permission from Elsevier.

siliciclastic environments can be ideal settings for the preservation of MISS, such as wrinkle structures. Figure 6.12 shows a variety of wrinkle structure morphologies which occur in different Paleoproterozoic sandstone packages from a highstand systems tract. Each of these wrinkle morphologies is interpreted to have been influenced in their development by the interaction of biotic processes of the microbial mat with physical sedimentary processes. As with stromatolites, wrinkle structure morphology can potentially be linked to different sedimentary environments.

Summary

Interest in stromatolites and MISS has been intense as, absent microbial body fossils, they are significant

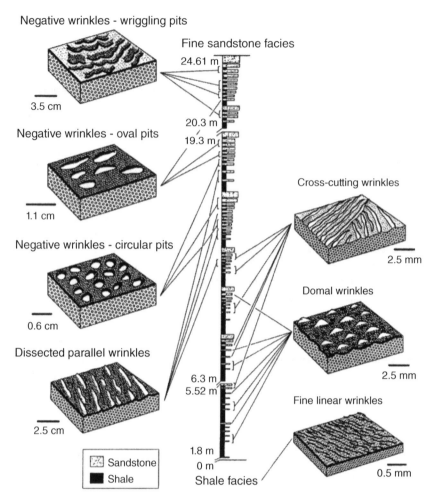

Figure 6.12 Distribution of different wrinkle structure morphologies through the lower part of a shallowing-upward highstand systems tract from the Paleoproterozoic Koldaha Shale in central India. These wrinkle structures are preserved abundantly on sandstone beds in these offshore marine environments. From Bannerjee and Jeevankumar (2005). Reproduced with permission from Elsevier.

indicators of the presence and behavior of microbes in ancient ecosystems. They are very important for understanding the ecology of the first 3 billion years of the history of life during the Precambrian before there were widespread metazoans. They also present important evidence for how early metazoans interacted with the microbial world ecologically and through evolution. In addition, these microbial structures have provided insight into ecosystem function after major mass extinctions. Advances in geomicrobiology have also led to the further understanding of the importance of microbes in the formation of other sediment components such as oolites (Summons et al. 2013). Study of microbes and their affects on sedimentary structures has also been spurred by the search for life in aqueous environments on other planets, such as Mars.

References

Andres, M.S. & Reid, R.P. 2006. Growth morphologies of modern marine stromatolites: A case study from Highborne Cay, Bahamas. *Sedimentary Geology* 185, 319–328.

Bailey, J.V., Corsetti, F.A., Bottjer, D.J. & Marenco, K.N. 2006. Microbially-mediated environmental influences on metazoan colonization of matground ecosystems: Evidence from the Lower Cambrian Harkless Formation. *Palaios* 21, 215–226.

Bannerjee, S. & Jeevankumar, S. 2005. Microbially originated wrinkle structures on sandstone and their stratigraphic context: Palaeoproterozoic Koldaha Shale, central India. *Sedimentary Geology* 176, 211–224.

Bose, S. & Chafetz, H.S. 2009. Topographic control on distribution of modern microbially induced sedimentary structures (MISS): A case study from Texas coast. *Sedimentary Geology* 213, 136–149.

Corsetti, F.A. & Grotzinger, J.P. 2005. Origin and significance of tube structures in Neoproterozoic post-glacial cap carbonates: Example from Noonday Dolomite, Death Valley, United States. *Palaios* 20, 348–362.

Gehling, J.G. & Droser, M.L. 2009. Textured organic surfaces associated with the Ediacara biota in South Australia. *Earth-Science Reviews* 96, 196–206.

Hagadorn, J.W. & Bottjer, D.J. 1997. Wrinkle structures: Microbially mediated sedimentary structures common in subtidal siliciclastic settings at the Proterozoic–Phanerozoic transition. *Geology* 25, 1047–1050.

Hagadorn, J.W. & Bottjer, D.J. 1999. Restriction of a Late Neoproterozoic biotope: Suspect-microbial structures and trace fossils at the Vendian–Cambrian transition. *Palaios* 14, 73–85.

Mariotti, G., Pruss, S.B., Perron, J.T. & Bosak, T. 2014. Microbial shaping of wrinkle structures. *Nature Geoscience* 7, 736–740.

Noffke, N. 2009. The criteria for the biogenicity of microbially induced sedimentary structures (MISS) in Archean and younger, sandy deposits. *Earth-Science Reviews* 96, 173–180.

Noffke, N. 2010. *Geobiology: Microbial Mats in Sandy Deposits from the Archean Era to Today*. Springer.

Oliver, L.K. & Rowland, S.M. 2002. Microbialite reefs at the close of the Proterozoic Eon: The Middle Member Deep Spring Formation at Mt. Dunfee, Nevada. *In* Corsetti, F.A. (ed.), *Proterozoic-Cambrian of the Great Basin and Beyond. The Pacific Section SEPM Book 93*, 97–118.

Shapiro, R.S. 2007. Stromatolites: 3.5-billion-year ichnologic record. In Miller, W.M., III (ed.), *Trace Fossils: Concepts, Problems, Prospects*. Elsevier, Amsterdam, pp.382–390.

Summons, R.E., Bird, L.R., Gillespie, A.L., Pruss, S.B., Roberts, M. & Sessions, A.L. 2013. Lipid biomarkers in ooids from different locations and ages: evidence for a common bacterial flora. *Geobiology* 11, 420–436.

Additional reading

Bailey, J.V., Orphan, V.J., Joye, S.M. & Corsetti, F.A. 2009. Chemotrophic microbial mats and their potential for preservation in the rock record. *Astrobiology* 9, 843–859.

Bottjer, D.J. 2005. Geobiology and the fossil record: Eukaryotes, microbes, and their interactions. *Palaeogeography, Palaeoclimatology, Palaeoecology* 219, 5–21.

Hagadorn, J.W., Pfluger, F. & Bottjer, D.J. (eds.). 1997. Unexplored microbial worlds. *Palaios* 14, 1–93.

Krumbein, W.E., Paterson, D.M. & Zavarzin, G.A. (eds.). 2003. *Fossil and Recent Biofilms: A Natural History of Life on Earth*. Kluwer Academic Publishers.

Reitner, J., Queric, N.-V. & Arp, G. (eds.). 2010. *Advances in Stromatolite Geobiology*. Lecture Notes in Earth Sciences Book 131, Springer.

Schieber, J., Bose, P.K., Eriksson, P.G., Banerjee, S., Sarkar, S., Altermann, W. & Catuneanu, O. (eds.). 2007. *Atlas of Microbial Mat Features Preserved Within the Siliciclastic Rock Record*. Elsevier.

7 Across the great divide: Precambrian to Phanerozoic paleoecology

Introduction

The Precambrian constitutes a great frontier for paleoecology. Because fossils in Archean and Proterozoic rocks typically are not of organisms with biomineralized skeletons, there is less of a fossil record than in the Phanerozoic. Because most of the Precambrian fossil record is microbial, much Precambrian paleoecology is understood in terms of microbial ecology. The predominant biogenic sedimentary structures of the Precambrian are stromatolites and microbially induced sedimentary structures (MISS), and a great deal has been learned from their study. Later in the Proterozoic, plants and protists evolved, followed by metazoans in the Neoproterozoic, and then at the end of the Neoproterozoic macrofossils and bioturbation first appeared. Actualism is applied toward understanding Precambrian biology and paleoecology, but this is also a time when a nonactualistic approach is most valuable. Because the fossil record is sparse and depends to a large extent on intervals with exceptional preservation, we only know the broad outlines of Precambrian paleoecology, in comparison to the Phanerozoic where the relatively rich fossil record has allowed a significant level of detail. During this time of transition into the Phanerozoic, the record of trace fossils and ichnofabric also plays a unique role in understanding ecosystem development.

Precambrian microbial paleoecology

There have been various attempts to understand microbial paleoecology for a number of intervals within the Precambrian. But the work that has been done has been confined to certain intervals, and there is much more that can be accomplished. Perhaps one of the most important problems in paleobiology is understanding the origin of life and the conditions under which early life flourished. In rocks approximately 3.5 billion years old, evidence for life on Earth becomes apparent. Some of this evidence includes various sorts of stromatolites such as seen in Fig. 7.1. These different stromatolite morphologies from the Archean Strelley Pool Formation in Western Australia were restricted to certain kinds of shallow carbonate environments. The distribution of stromatolite morphologies is considered to be evidence that they were primary biological structures and not abiogenic structures. Such evidence tells us that microbial life living in mats was forming microbial carbonates and ecosystems on the seafloor in many ways similar to what we see through the rest of the Precambrian, where stromatolites are also common to abundant.

Paleoecology: Past, Present and Future, First Edition. David J. Bottjer.
© 2016 John Wiley & Sons, Ltd. Published 2016 by John Wiley & Sons, Ltd.

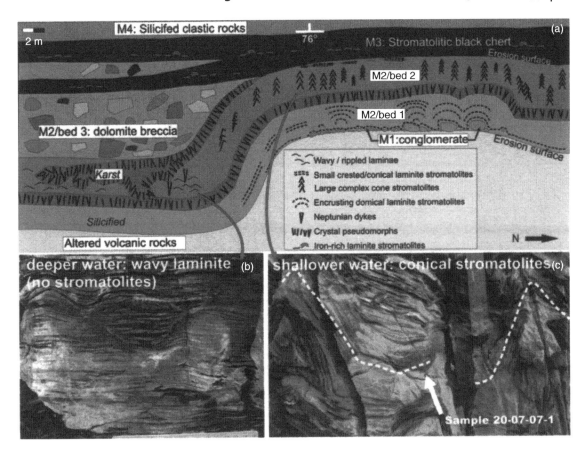

Figure 7.1 Archean (~3.45 billion years old) reef-like assemblage of stromatolites from the Strelley Pool Formation at the platform margin outcrop on southern Trendall Ridge (Western Australia). (a) Outcrop map showing cross-sectional view of stratigraphy from underlying altered volcanic rocks up through members (M) 1–3 and part of member 4 of the Strelley Pool Formation. Of special interest is the paleotopographic feature, with stromatolites only deposited on the high side (right). (b) Wavy laminites deposited in deeper water south (left) of the paleohigh. (c) Large complex conical stromatolites formed on the paleohigh. The dotted white line traces a single lamina across two coniform stromatolites. Scale rule in B and C is 15 cm. From Allwood et al. (2009). Reproduced with permission from the National Academy of Sciences. (*See insert for color representation.*)

With this evidence of microbial mats and their products in carbonate environments, it has become apparent that siliciclastic environments should also be searched for the presence of MISS, so that the microbial world of the Precambrian and its ecology can be even better understood. The presence of MISS in various sedimentary units has been documented in recent years. For example, a study of a 2.9 billion-year-old tidal flat complex from South Africa (Fig. 7.2) shows evidence for deposition of a variety of MISS, including erosional remnants and pockets, oscillation cracks, gas domes, and multidirected ripple marks. This evidence indicates that microbial mats were common and that microbial ecology was complex in these sedimentary environments.

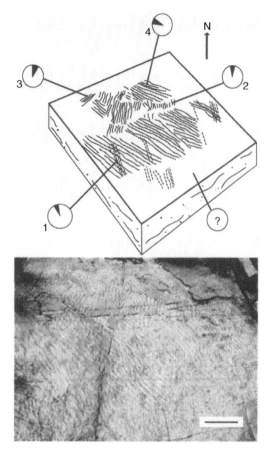

Figure 7.2 Archean (2.9 billion years old) microbially induced sedimentary structures (MISS) from the Nhlazatse Section, Pongola Supergroup, South Africa. These multidirected ripple marks which were deposited in tidal flat settings are exposed on a 7–10 cm thick fine sandstone bed; scale: 1 m. The ripple pattern is composed of patches (or groups) of ripple marks. Each patch (or group) is characterized by one ripple crest direction as shown in the stereonets. The analysis of the pattern shows that the ripple mark groups were formed subsequently during individual reworking events. Between reworking events and formation of ripples, the substrate was stabilized by development of a cyanobacterial microbial mat. The numbers 1–4 indicate the order of formation of the ripple generations. The first generation ripple marks are least visible and mostly occupy the smallest area on the bedding plane: (1) 5%; (2) about 1.5%; (3) 17%; (4) 45%; (nonpreserved) 33%. From Noffke et al. (2008). Reproduced with permission from John Wiley & Sons. (*See insert for color representation.*)

Early animals

While there remains much to be done with Precambrian microbial paleoecology, fossils of early animals are at times enigmatic, and their phylogenetic affinities are commonly difficult to interpret in terms of modern animals on Earth. Such determinations are essential for understanding the ecosystems in which these early animals evolved. Phylogenetics has been revolutionized over the past few decades with the use of cladistic methodology, which determines monophyletic groups through their possession of shared derived characters. With the advent of cladistics in paleontology, one of the most valuable concepts in understanding early animal evolution is that of the stem group and the crown group (Brysse 2008). The crown group for a phylum includes all the representatives of that phylum and their evolutionary characters that we find on Earth today as well as their ancestors with those characters. In contrast, the stem group of a phylum includes the extinct members which do not have all of the characters that we find in the crown group. For example, crown group arthropods include a molting cuticle, a chitinous exoskeleton, a segmented body, and jointed appendages. But there are members of the arthropods that do not include jointed appendages, and these are extinct and are the stem arthropods. The crown and stem group concept provides an insightful analytical tool toward understanding the evolutionary relationships of early animals and also allows a better understanding of the paleoecology of these organisms.

It has generally been thought that, as oxygen content in the oceans increased, animal evolution became possible. Evidence for early animals from the stratigraphic record goes back to before the Ediacaran to the end of the Cryogenian, which occurred some 635 million years ago, in the form of sponge biomarkers. Other sorts of evidence for early animals, including sponges include the phosphatized microfossils of the Doushantuo Formation in China, likely 600 million years old (Fig. 7.3). Discussions on the nature of these fossils, including whether they might represent stem or crown group animals,

Figure 7.3 Scanning electron micrograph of a Doushantuo Formation (China) phosphatized microfossil that is interpreted to be a sponge grade body fossil with cellular preservation. This image shows the main tubular chamber of the specimen with a large opening and additional chambers, viewed from the exterior. Cellular preservation is shown here on the surface which is covered with cells closely resembling sponge pinacocytes. From Yin et al. (2015). Reproduced with permission from the author.

are ongoing, and further information is required before definitive paleoecological interpretations of Doushantuo microfossils can be made.

In the fossil record, the first great animal biota with a fairly extensive fossil record is the Ediacara biota, which occurs in the Ediacaran from approximately 575 million years ago to just before the Cambrian, which began 542 million years ago. Fossils of the Ediacara biota are macroscopic, and their large size was presumably an evolutionary response to increasing oxygen levels. These fossils are preserved due to microbial sealing and obrution events as discussed earlier. Ediacara biota fossils are found preserved in a variety of environments from the shelf (Fig. 7.4) to the deep sea (Fig. 4.14). These animals had originally been interpreted as crown group members of current phyla, but now, it is thought that although some of them may represent extinct groups (e.g., Fig. 4.14), many of them are probably stem group representatives of Porifera,

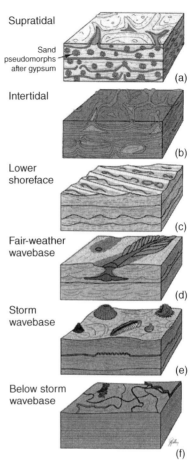

Figure 7.4 Typical shoreline and shelf environments of the Rawnsley Quartzite of South Australia showing occurrence of Ediacara biota fossils from fair-weather wave base and deeper. Depth-related associations of microbial mats and fossils are (a) supratidal sandflats with desiccation of microbial mat-bound sand lamination and sand-pseudomorphs after sulfate; (b) intertidal sandflats with domed and disrupted microbial mats; (c) lower shore-face wave rippled sand with syneresis cracks in microbial-bound ripple troughs; (d) fair-weather wave base with preservation of anchoring holdfasts of fronds, either flattened or torn away by storm surge; (e) storm wave base with communities of benthic Ediacara organisms; and (f) below wave base preservation, below distal flows of storm surge sand. From Gehling and Droser (2012). Reproduced with permission from the International Union of Geological Sciences.

Figure 7.5 Diorama depicting the Ediacara biota from South Australia, including a large *Dickinsonia* (center top), *Spriggina* (lower right corner), *Kimberella* (lower left center), and *Tribrachidium* (lower left corner), as well as a variety of fronds. From Zimmer (2006). (*See insert for color representation.*)

Cnidaria, Placozoa, and various bilaterians. Other animals also have their first appearance in the fossil record in the Ediacara biota. Figure 7.5 shows a diorama of the Ediacara biota community from the Ediacara Hills of Australia showing a number of enigmatic animals. In the lower left of this diagram is illustrated *Kimberella*, which is thought to be a stem group mollusc that moved about on the surface and scratched microbial mats for its food resources. Perhaps one of the most enigmatic of the Ediacara fossils is *Dickinsonia*, the large fossil shown in the center left, which grew to as much as a meter in greatest dimension. The interpretations of the biological affinities of *Dickinsonia* have been wide ranging. However, recent interpretation by Erik Sperling and Jakob Vinther of *Dickinsonia* as a stem group placozoan has great appeal. In this interpretation, allying *Dickinsonia* with the Placozoa is insightful because *Dickinsonia* does not have a mouth, an anus, or an apparent digestive system. The modern Placozoa use extracellular digestion of

materials along the bottom or sole of the animal, and this is the mode of nutrition proposed for *Dickinsonia*. There are trace fossils of *Dickinsonia* that show it apparently moved and then stayed in one place for a while so that if it was a stem group placozoan, it might have been digesting the microbial mat. Thus, this was a mat-based world where a number of organisms lived on mat trophic resources. This is notable in that microbial mats played such a large role not only in the ecology and morphology of a number of organisms of the Ediacara biota but also, as discussed in Chapter 4, their preservation (Figs. 4.12 and 4.13), since they are all soft-bodied organisms. There also are frond-shaped organisms that were probably engaged in some sort of suspension feeding and attest to other sorts of food resources in the water column (Figs. 4.13 and 4.14). The large size of organisms in the Ediacara biota has been hypothesized to have developed because macropredation had not evolved by this time (Fig. 7.11).

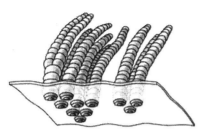

Figure 7.6 Reconstruction of *Funisia dorothea* from South Australia with holdfast beneath mat substrate. Each individual organism is up to 30 cm long. From Droser and Gehling (2008). Reproduced with permission from the American Association for the Advancement of Science.

The number and types of organisms in the Ediacara biota are not completely known and are continually being expanded, thus improving our understanding of how Ediacaran ecosystems operated. For example, recently, a new type of organism, *Funisia*, which is thought to be one of the more common organisms in the South Australian Ediacara biota, was described and interpreted as a stem group sponge. This organism is fairly simple in morphology and is illustrated in Fig. 7.6.

Work is ongoing to understand the paleoecology of the Ediacara biota. In general, the Ediacara biota was an epifaunal community of organisms, and these characteristics have been incorporated into overall tiering histories (Fig. 1.3). However, at a finer temporal scale, it is possible to track changes in tiering in the Ediacara biota through time. Figure 7.7 shows this sort of analysis where the diorama in Fig. 7.7A represents the oldest Ediacara community from Mistaken Point, Newfoundland, and the diorama in Fig. 7.7C represents one of the youngest Ediacara faunas from Namibia. One of the trends that has been documented for the Ediacara biota is that the addition of fossils that appear to represent mobile bilaterian animals occurs as the biota becomes younger. This diagram shows that the amount of epifaunal tiering also appears to have decreased through time from the oldest to the youngest community of the Ediacara biota. This decrease has been attributed to the increased sediment disturbance that would develop due to the

presence of increasing numbers of mobile seafloor organisms, which would have created unstable conditions for sessile epifaunal organisms.

Paleoecology of the Cambrian fauna

In the latest Ediacaran and early Cambrian, there appear small mineralized fossils, primarily micro-fossils, called the "small shelly fauna". This includes the latest Ediacaran *Cloudina*, a conotubular organism of unknown phylogenetic affinity which was adapted to living in benthic microbial mat conditions (Fig. 7.8). Most of the small shelly fauna comprises the skeletal parts of larger organisms. A variety of explanations have been offered for this initiation of biomineralization, but one of the compelling hypotheses is that this represents an adaptive response to the initial evolution of macropredation. These small shellies are then joined in the early Cambrian by trilobites and other larger organisms with mineralized skeletons and this time largely coincides with the beginning of the "Cambrian explosion." The Cambrian explosion received its name because most of the readily fossilized phyla have their first appearance in the Cambrian period. The phyla with a later appearance in the fossil record tend to be those that do not fossilize well, including many that have no fossil record because they have apparently very poor fossil preservation potential. With the Cambrian explosion, paleoecology begins to resemble that for the rest of the Phanerozoic, with preservation of a prolific amount of skeletonized fossils to provide information on ancient ecology.

Another compelling feature of the fossil record that relates to ecology in a dramatic way is the increase in size of organisms from the Ediacaran into the Phanerozoic as compiled by Jonathan Payne and colleagues (Fig. 7.9). The appearance of the earliest plant *Grypania* in the Paleoproterozoic represented a very big increase in size, and the next big increase was not until the evolution of *Dickinsonia* from the Ediacara biota. Size increase continues into the Cambrian with the top predator in marine environments, *Anomalocaris*, which was

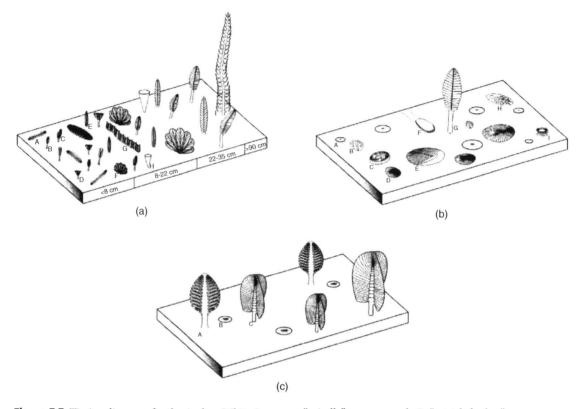

Figure 7.7 Tiering dioramas for the Avalon, White Sea, and Nama Assemblages. The Avalon Assemblage (575–560 million years ago), typified by occurrences at Mistaken Point (Newfoundland, Canada), lived in deep-sea environments. The White Sea Assemblage (560–550 million years ago) is represented by fossils deposited in shallow marine environments from the White Sea region in Russia as well as from South Australia. The Nama Assemblage (549–542 million years ago), represented by fossils from Namibia, are from very shallow marine environments. The dioramas are composites of a number of individual communities and do not intend to show the proportion of organisms in each tier. (a) Avalon Assemblage. Fossils depicted: A, "spindle" rangeomorph; B, "ostrich feather" rangeomorph; C, *Charniodiscus*; D, "feather duster" rangeomorph; E, *Charnia* rangeomorph; F, *Bradgatia* rangeomorph; G, "pectinate" rangeomorph; H, *Thectardis*; I, "xmas tree." (b) White Sea Assemblage. Fossils depicted: A, discoidal fossil (probable holdfast); B, *Parvancorina*; C, *Tribrachidium*; D, *Dickinsonia*; E, *Yorgia* with grazing impression; F, *Kimberella* with *Radulichnus* trace; G, *Charniodiscus*; H, Vendomiid (e.g., *Vendia*); I, *Eoporpita*. (c) Nama Assemblage. Fossils depicted: A, *Rangea*; B, *Nemiana*; C, *Swartpuntia*. From Bottjer and Clapham (2006). Reproduced with permission from Springer Science and Business Media.

a long as a meter, and then continues on up to the modern. So the Cambrian explosion not only had many new organisms on a macroscale appearing to make more complex ecology, but it was also part of a trend in which organisms became larger and larger, attributed to increasing oxygenation levels (Fig. 7.9).

Along with this increase in organism size, the ability to burrow below the surface of marine sediments was also evolving in the beginning of the Phanerozoic. The use of the ichnofabric index approach (Figs. 5.4 and 5.5) shows that the amount of reworking by bioturbation recorded in Cambrian sediments increased dramatically from the earliest

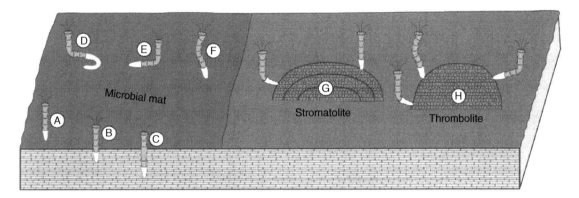

Figure 7.8 Reconstruction of *Cloudina* life modes in a mat-based benthic environment. Tubes of modern serpulid worms are thought to be the most appropriate modern analogue for paleoecological interpretations. Variety of morphology (A-H) for this conotubular organism is interpreted to be caused by a sessile suspension-feeding life style, where the organism was living on or inserted into (matsticker; Fig. 7.11) level-bottom microbial mats, or on stromatolites and thrombolites. *Cloudina* is usually found as a millimeter-scale fossil, although some at the centimeter scale occur; for the purposes of clear presentation the sizes of *Cloudina* in the schematic have been significantly enlarged. From Cai et al. (2014), where data sources are indicated. Reproduced with permission from Elsevier.

Cambrian to later Cambrian times and increased again at the end of the Ordovician (Fig. 7.10). This increase in average ichnofabric indicates increased ecological complexity from the Ediacaran into the Phanerozoic. The dramatic increase in reworking of sediments has been termed the agronomic revolution by Dolf Seilacher and Friedrich Pflüger. The overall pattern of the agronomic revolution is illustrated in Fig. 7.11, where on the left the microbial mat-dominated environment and ecology of the Ediacara biota with Precambrian matgrounds is depicted and on the right is a seafloor which developed in the Phanerozoic as a well-developed surface mixed layer ("mixgrounds"). The agronomic revolution affected the morphologies and lifestyles of organisms through this interval, and many of them were adapted to living on a microbial mat-covered seafloor, which was followed after the Ediacaran by evolution of organisms living on the seafloor on or within bioturbated sediment.

This increase in bioturbation was not a uniform process that happened in all environments at the same time. Rather much of the Cambrian seafloor consisted of a mosaic of microbial mat-dominated seafloors and more bioturbated seafloors. A study of lower Cambrian wrinkle structure distribution across environments shows that these were best preserved in the offshore transition as well as in tidal flat settings (Fig. 7.12). Thus, in the Cambrian are found organisms which were adapted to microbial mat environments as well as those that were adapted to bioturbated environments. In work with James Hagadorn and Stephen Dornbos, this pattern has been termed the Cambrian substrate revolution, which is further illustrated by examining the change in processes that affected subtidal sediment deposition as seen in Fig. 7.13. At the end of the Neoproterozoic (Ediacaran), sediments were primarily influenced by physical and microbial processes. But, as previously discussed, sediments in the Cambrian were dominated by physical and microbial processes as well as metazoan processes, which are those which cause disruption by bioturbation. After the Cambrian, it is largely the case that a switch had been made to the mode whereby bioturbation was much more significant than microbial processes for formation of sedimentary structures. From there on, physical and metazoan processes dominated sedimentary fabric in most depositional systems (Fig. 7.13).

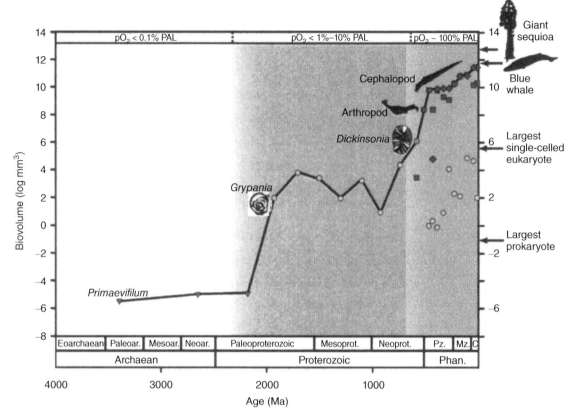

Figure 7.9 Change in size of largest fossils through Earth history. Size maxima are plotted as computed biovolume (log mm³). Single-celled eukaryotes, animals, and vascular plants are illustrated separately for the Ediacaran and Phanerozoic. The solid line denotes the trend in the overall maximum for all of life. Increases in the overall maximum occurred in discrete steps approximately corresponding to increases in atmospheric oxygen levels in the mid-Paleoproterozoic and Ediacaran–Cambrian–early Ordovician. Sizes of the largest fossil prokaryotes were not compiled past 1.0 Gya. Estimates of oxygen levels are expressed in percentages of the present atmospheric level (PAL). Phan., Phanerozoic; Pz, Paleozoic; Mz, Mesozoic; C, Cenozoic. Triangles, prokaryotes; circles, protists; black squares, animals; diamonds, vascular plants; gray square, probable multicellular eukaryote. From Payne et al. (2009). Reproduced with permission from the National Academy of Sciences. (*See insert for color representation.*)

Thus, if one examines Cambrian seafloor communities and how they lived, it is apparent that many are characterized by epifauna. For example, Fig. 7.14 represents an example from the early Cambrian Chengjiang biota in China where there is relatively little bioturbation (Fig. 4.11) and the organisms lived attached to or moved around on the surface of the seafloor. Chengjiang biota communities can include large numbers of small priapulid worms, but these relatively immobile infaunal organisms contributed little to the development of a surface mixed layer (Dornbos and Chen 2008). Thus, many Chengjiang biota organisms retained an ecology that is fundamentally driven by the low level of bioturbation on the seafloor.

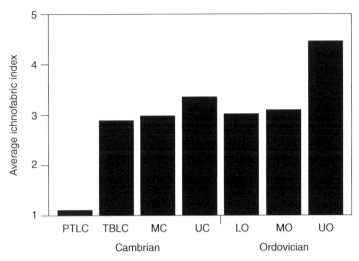

Figure 7.10 Increase in bioturbation through the Cambrian and Ordovician portrayed as average ichnofabric index from carbonate inner shelf paleoenvironments for Cambrian and Ordovician strata of the Great Basin, United States. PTLC, pre-trilobite lower Cambrian; TBLC, trilobite-bearing Cambrian; MC, Middle Cambrian; UC, Upper Cambrian; LO, Lower Ordovician; MO, Middle Ordovician; UO, Upper Ordovician. From Droser and Bottjer (1993), where data sources are indicated.

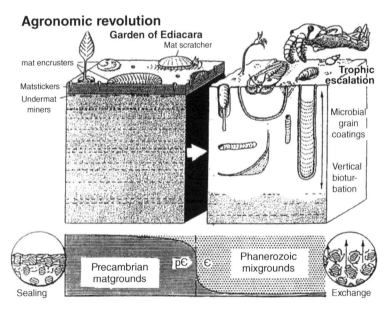

Figure 7.11 The typical presence on seafloors of microbial mats, or matgrounds, during the Precambrian was changed by the evolution of vertical bioturbation and the development of mixgrounds, and this transition has been termed the agronomic revolution. The Phanerozoic also saw the evolution of macropredation, and this has led to the concept of the Garden of Ediacara, before such predation existed. Life modes for animals in the Ediacara biota can be categorized as follows: mat scratchers are organisms that graze the surface of microbial mats, mat encrusters are epifaunal organisms that attach to the surface of microbial mats, matstickers are organisms that are semi-infaunal and are inserted into the microbial mat, and undermat miners are burrowing organisms that burrow below the surface mat. From Seilacher (1999). Reproduced with permission from the SEPM Society for Sedimentary Geology.

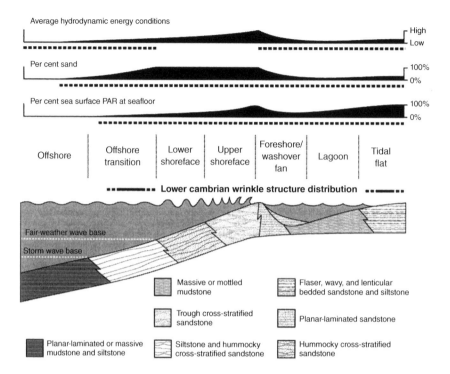

Figure 7.12 In the Cambrian, wrinkle structures were preferentially preserved in some environments. This facies model shows the preferred environments for wrinkle structure formation and preservation (marked by dashed lines) based upon the parameters of average hydrodynamic energy conditions, percent sand, and percent sea-surface photosynthetically active radiation (PAR) at the seafloor. If most wrinkle structures are formed by phototrophic microbial mats, this would limit most wrinkle structures to above the euphotic depth, which is approximated by 1% PAR. Wrinkle structures show a preference for being preserved atop quartz sandstones, and this would cause wrinkle structure preservation in environments where sand can be delivered by storm waves, currents, and tidal activity. This would limit wrinkle structures to environments primarily above storm wave base. Also, wrinkle structures should be prohibited from high-energy environments like the shoreface, because flow velocities are strong enough to erode an incipient mat. With these restrictions, this leaves the offshore transition, lagoon, and tidal flats as the primary environments in which wrinkle structures should be preferentially preserved. Lower Cambrian wrinkle structures are found primarily within the proximal portion of the offshore transition and within midtidal flat settings. From Mata and Bottjer (2012), where data sources are indicated. Reproduced with permission from Springer Science and Business Media.

Examination of other environments in the Cambrian, such as the deeper-water settings where the Burgess Shale was deposited, also shows evidence, through the study of body fossils, of limited bioturbation by organisms including priapulid worms (Fig. 7.15). Further studies of this important lagerstätten have revealed more about the role of Burgess Shale substrates in the ecology of Cambrian explosion organisms. With an understanding of the importance of mat-based ecology for the Cambrian, several important new ecological interpretations have been made. For example, the Burgess Shale fossil *Odontogriphus* has been interpreted as a stem group mollusc which crawled over the seafloor and ingested microbial mats for its nutrition. A synthesis of early mollusc evolution and ecology from the

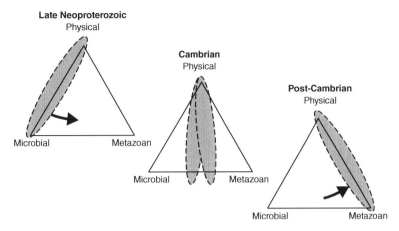

Figure 7.13 Change in dominant processes controlling the sedimentary fabric of normal marine neritic soft substrates during the late Neoproterozoic (Ediacaran)–Phanerozoic transition. These processes are primary physical and microbial processes, as well as secondary metazoan bioturbation. The movement of the ovals within the triangle indicates changes in the relative dominance of these processes. During this transition, the dominant processes move from a combination of physical and microbial processes to a combination of physical and metazoan processes. The phenomenon of benthic metazoan extinction and adaptations to these transitional soft substrates has been termed the Cambrian substrate revolution. From Dornbos et al. (2005). Reproduced with permission from Elsevier.

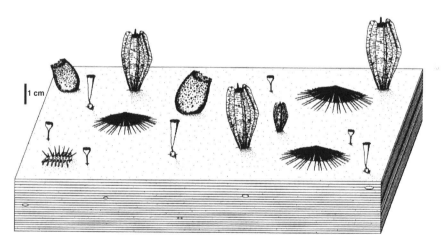

Figure 7.14 Life habits of Chengjiang biota (China) benthic suspension feeders living on a substrate with very little bioturbation. This diorama shows Maotianshan Shale benthic suspension-feeding genera in life position on top of a substrate with just a few horizontal burrows seen as oval cross sections a few millimeters in diameter. An example of the sediment resting demosponge *Crumillospongia* is in the upper left corner. An example of the shallow sediment sticking demosponge *Takakkawia* is in the upper right corner. An example of the sediment resting demosponge *Choia* is next to this *Takakkawia*. An example of the hard substrate attaching possible cnidarian *Cambrorhytium* is in the lower right corner. Examples of the enigmatic shallow sediment sticker *Dinomischus* are in the lower left corner next to the lobopodian *Hallucigenia*. From Dornbos et al. (2005). Reproduced with permission from Elsevier.

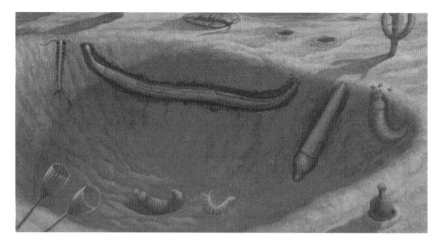

Figure 7.15 Schematic diagram of Burgess Shale infauna envisaged as if a modern submersible was there and had scooped a sample of seafloor sediment. The infauna is dominated by priapulid worms, of which the most abundant was *Ottoia*. In this scene, three individuals are visible; one on the floor of the large excavation, another in the process of consuming hyoliths (mid-right), while the third is emerging from its burrow and displaying its spinose proboscis (lower right). Two other priapulids are visible in the excavation: the elongate, more-or-less horizontal worm is *Louisella*, shown here in its life position as a sedentary animal occupying an elongate burrow with openings to the overlying seawater at either end. The animal inclined downward, with its posterior end just emerging from the seafloor, is an example of *Selkirkia*. It inhabited a parchment-like tube and in common with other priapulids had a spiny proboscis that was employed, when necessary, for burrowing. The other types of worm visible in the excavation are two examples of the polychaete annelid *Burgessochaeta*, with one individual wriggling on the floor and the other in its burrow with anterior tentacles extending sideways (far left). Above the seafloor are a typical Burgess Shale sessile epifauna – a trilobite (top), the sponge *Vauxia* (upper right), and the enigmatic *Dinomischus* (lower left). From Conway Morris (1998). Reproduced with permission from Oxford University Press. (*See insert for color representation.*)

Ediacaran into the Ordovician by Jean-Bernard Caron and others is shown in Fig. 7.16. This synthesis presents early molluscs typified by *Kimberella* from the Ediacara biota as well as *Odontogriphus* and *Wiwaxia* and other stem group molluscs as influenced primarily by the mat-based ecology and shows that they moved over the surface of the mat and ingested it in various ways. With further increase in bioturbation this mat-based ecology disappeared and so did these early molluscs, although as shown in Fig. 7.16 some crown group molluscs such as the chitons (Polyplacophora), which evolved a little later in the Cambrian, also lived by scraping mats on various soft and hard substrates.

These sorts of trends can also be seen in the evolutionary and ecological history of early sessile suspension-feeding stem group echinoderms (Fig. 7.17). Among the early echinoderms the helicoplacoids, which were shaped like a top, were perhaps the strangest. These suspension feeders were inserted into relatively stiff seafloor sediment that was bound by microbes. Helicoplacoids became extinct later in the Cambrian; their extinction is interpreted to have occurred because their environments were more thoroughly bioturbated, and thus, they lost their ability to be stable on the seafloor. In contrast, the edrioasteroids, which lived on microbially bound soft sediments in the earliest part of the Cambrian, evolved in response to increasing levels of disturbance by bioturbation through attachment to hard objects on the seafloor such as shells or rocks. Similarly, eocrinoids, which

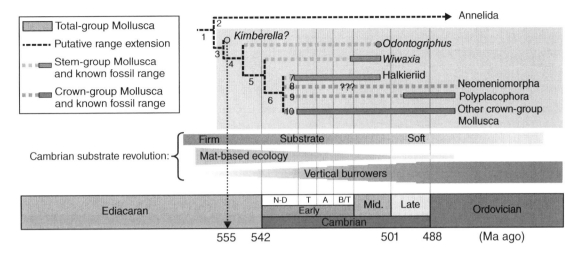

Figure 7.16 Evolution of the molluscs and the Cambrian substrate revolution. A proposed evolutionary tree for molluscs indicates this sequence of evolution of various characters during the Ediacaran and Cambrian. (1) Protostome bilaterians; serial replication; triploblastic. (2) Segmentation by coelomic metameres. (3) Large size; with iteration but not coelomic segmentation; ovoid; dorsoventrally flattened; stiffened cuticular dorsum; flat, noncuticularized ventral sole; radula of iterated, paired mirror-image teeth and radular membrane (certain for *Odontogriphus*); feeding on biomat? (4) Groove (mantle cavity) between dorsum and ventrum with serial ctenidia; paired salivary glands; straight digestive tract; nervous system ladderlike?; coelom posterior; restricted to reproductive and excretory organs? (5) Noncalcified scleritome, sclerites arranged in three mirror-image longitudinal zones. (6) Calcification of epidermally nucleated sclerites that pass through the cuticle; calcified shell from serial shell fields; no periostracum from periostracal groove of mantle lobe. (7) Two shell fields. (8) Tubiform; reduced foot; sclerites in 1–3 longitudinal rows beside foot groove; progenetic loss of gills and shells; embryological evidence of vestigial shell fields. (9) Eight or more shell fields; sclerites not in longitudinal zones. (10) Loss of sclerites and serial shell fields; true periostracum secreted from mantle lobe; shells paired or single, reduction of gills; further variety of body plans. A, Atdabanian; B/T, Botomian/Toyonian; N-D, Nemakit-Daldynian; T, Tommotian. From Caron et al. (2006). Reproduced with permission from Macmillan Publishers Ltd. (*See insert for color representation.*)

earlier in the Cambrian also lived on microbially bound seafloors, with increasing bioturbation later evolved a stalk and attached to hard surfaces.

This analysis for helicoplacoids, edrioasteroids, and eocrinoids has been extended through their entire fossil records (Fig. 7.18). Most of these animals that lived in the Cambrian were adapted for living on more stable substrates, typically characterized by a microbial mat, that were typical for the Proterozoic. As these Proterozoic-style substrates disappeared due to the continued evolution of more extensive bioturbation, a Phanerozoic-style substrate with development of a surface mixed layer due to bioturbation became predominant. Thus, after the Cambrian, these organisms primarily exhibit adaptations to Phanerozoic-style substrates, before their complete extinction by the end of the Carboniferous. These trends and adaptations from the more microbially dominated seafloors of the early and Middle Cambrian to those later in the Paleozoic are illustrated in Fig. 7.19.

Another stem group of echinoderms is the stylophorans, which also range from the Cambrian to the Carboniferous (Fig. 7.20). Stylophorans appear to have been somewhat mobile but lay flat on or partially buried in the seafloor, with several

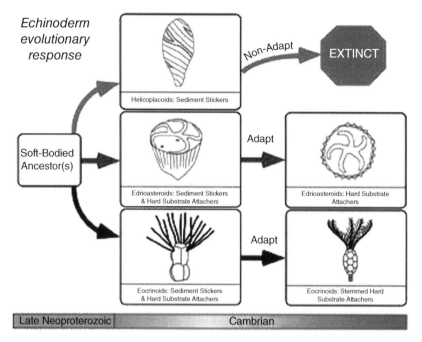

Figure 7.17 Evolution of the echinoderms and the Cambrian substrate revolution. Arrows do not represent a direct evolutionary relationship between specific echinoderms shown, but imply a general evolutionary trend through the Cambrian within each of the groups examined, with these echinoderms serving as individual examples. Helicoplacoid drawing is from specimen 3 cm in height. For edrioasteroids, *Camptostroma* (left) drawing is from specimen 5 cm in height. Edrioasteroid on right is a schematic of a typical attaching edrioasteroid with width size of 5 cm. For eocrinoids, *Lichenoides* (left) drawing is from a specimen 2.5 cm in height. Eocrinoid on right is *Tatonkacystis*, drawing is from specimen approximately 5 cm in height. Geological time not to scale and boxes do not represent the precise age range of the echinoderms they contain. From Bottjer et al. (2000). Reproduced with permission from the Geological Society of America. (*See insert for color representation.*)

protuberances anchoring them to the soft substrate (Fig. 7.20). They typically directed their one arm, which was a suspension-feeding structure, into the current. Because largely sessile stylophorans were dependent upon seafloor stability, they are interpreted by Bertrand Lefebvre to have become extinct later in the Paleozoic because of decreased seafloor stability due to increased bioturbation. Thus, they are likely to have also been an extended part of the phenomenon of the Cambrian substrate revolution. These stem group echinoderms were an integral part of the Cambrian evolutionary fauna (Figs. 3.10 and 3.12), while the crown group echinoderms, the echinoids, holothurians, ophiuroids, asteroids, and crinoids, all evolved mobility or, in the case of the stalked crinoids, holdfast structures, that allowed them to thrive on Phanerozoic-style soft substrates.

Summary

The analysis of paleoecology during the Precambrian to Phanerozoic transition has moved forward with the steadily increasing understanding that has been made on the nature of the organisms which make up the fossil record during this time. The appreciation of the stem group concept has gone a

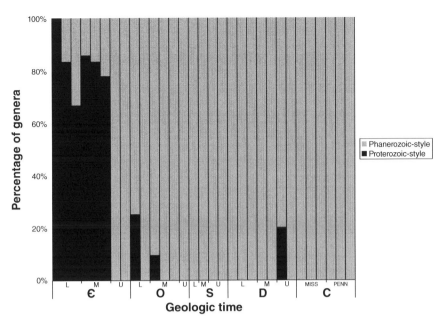

Figure 7.18 The Phanerozoic fossil record of three epifaunal sessile suspension-feeding stem echinoderm groups (helicoplacoids, edrioasteroids, eocrinoids) that originated in the Cambrian and the Cambrian substrate revolution. The stacked-bar histogram shows percentage of genera for these three groups within each stage of the Cambrian through the Carboniferous and whether they were adapted to typical Proterozoic-style firm soft substrates (black) without a bioturbated mixed layer, or typical Phanerozoic-style soft substrates (gray), with a bioturbated mixed layer. Adaptations to Proterozoic-style soft substrates include sediment resters, shallow sediment stickers, and sediment attachers, similar to life modes outlined for the Ediacara biota (Fig. 7.11). Adaptations to Phanerozoic-style soft substrates include snowshoe strategists, which have a broad basal surface for stability on the surface; iceberg strategists, which submerge a significant portion of the animal below the seafloor for stability; holdfast strategists, which have evolved specialized holdfasts for stability on soft substrates; and hard substrate attachers. Sample size for each bar of the histogram, from left to right: 2, 6, 3, 7, 6, 9, 2, 6, 4, 8, 11, 8, 25, 16, 4, 6, 5, 2, 2, 2, 3, 2, 4, 5, 3, 7, 7, 6, 2, 4, 2. The two initial stages (Nemakit-Daldynian and Tommotian) and final stage (Trempealeauan) of the Cambrian contain no data and thus are not included in this histogram. From Dornbos (2006). Reproduced with permission from Elsevier.

long way toward clarifying our understanding of the phylogenetic affinities of many organisms (Brysse 2008). In particular, fossils of the Ediacara biota and the Doushantuo biota are still under discussion as to their phylogenetic affinities (e.g., Retallack 2013; Xiao et al. 2013; Chen et al. 2014; Yin et al. 2015). Recent advances in understanding paleoecological structure have developed from a clearer understanding of the record of microbial structures as well as trace fossils and ichnofabric during the Neoproterozoic to Phanerozoic transition (e.g., Tarhan and Droser 2014). Thus, it is apparent that evolutionary paleoecology of many Ediacaran and early Paleozoic benthic organisms was influenced greatly by the nature of the seafloor. This is best seen in molluscs and echinoderms, which show a broad range of adaptations and extinctions reflecting the changing ecological landscape as organisms evolved to burrow deeper and bioturbate at greater rates. A variety of other ecosystem parameters, such as

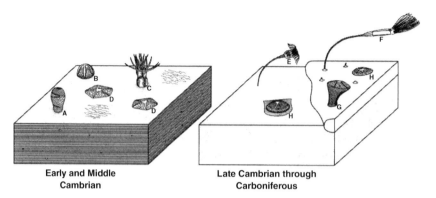

Early and Middle Cambrian **Late Cambrian through Carboniferous**

Figure 7.19 The Cambrian substrate revolution and sessile suspension-feeding stem echinoderms. Dioramas depicting dominant substrate adaptations of sessile benthic suspension-feeding echinoderms and dominant soft substrate characteristics during early and Middle Cambrian (left) and from Late Cambrian through Carboniferous (right). Left box includes (A) shallow sediment sticker, the helicoplacoid *Helicoplacus*; (B) sediment attacher, the edrioasteroid *Totiglobus*; (C) shallow sediment sticker, the eocrinoid *Lichenoides*; and (D) sediment attacher, the edrioasteroid *Stromatocystites*. Note low levels of horizontal bioturbation and wrinkle structures on microbial mat-bound substrate. Right box includes (E) holdfast strategist, the eocrinoid *Macrocystella*; (F) hard substrate attacher, the eocrinoid *Haimacystis*; (G) hard substrate attacher, the edrioasteroid *Hypsiclavus*; and (H) generalized hard substrate attaching pyrgate edrioasteroids. Note well-bioturbated soft sediment and attachment to skeletal material and carbonate hardgrounds (right). Box diagrams not intended to show echinoderms that lived at the same time or all possible substrate adaptations, just dominant substrate adaptations during specific time intervals. From Dornbos (2006). Reproduced with permission from Elsevier.

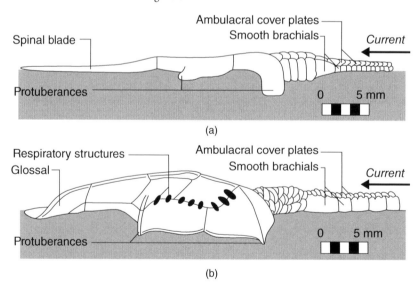

Figure 7.20 Morphology and life position of stylophorans, a stem group of echinoderms that ranges from the Cambrian to the Carboniferous (in right lateral view). A, *Cothurnocystis elizae*, Starfish Bed (Upper Ordovician), Ayrshire (Scotland). B, *Ceratocystis perneri*, Skryje Shales (Middle Cambrian), Czech Republic. Brachials are arm plates, ambulacral plates are part of the water vascular system, protuberances anchor the animal in the substrate, and the shaded area represents seafloor sediment. From Lefebvre (2003). Reproduced with permission from John Wiley & Sons.

the evolution of reefs (Figs. 9.5 and 12.17) and predation (e.g., Erwin et al. 2011), were also driving the structure of these early ecosystems. Much remains to be done, and this interval in life's history has many exciting opportunities for the student of evolutionary paleoecology.

References

Allwood, A.C., Grotzinger, J.P., Knoll, A.H., Burch, I.W., Anderson, M.S., Coleman, M.L. & Kanik, I. 2009. Controls on development and diversity of Early Archean stromatolites. *Proceedings of the National Academy of Sciences* 106, 9548–9555.

Bottjer, D.J. & Clapham, M.E. 2006. Evolutionary Paleoecology of Ediacaran Benthic Marine Animals. *In* Xiao, S. & Kaufman, A.J. (eds.), *Neoproterozoic Geobiology and Paleobiology*, Springer, pp. 91–114.

Bottjer, D.J., Hagadorn, J.W. & Dornbos, S.Q. 2000. The Cambrian substrate revolution. *GSA Today* 10, 1–7.

Brysse, K. 2008. From weird wonders to stem lineages: The second reclassification of the Burgess Shale fauna. *Studies in History and Philosophy of Biological and Biomedical Sciences* 39, 298–313.

Cai, Y., Hua, H., Schiffbauer, J.D., Sun, B. & Yuan, X. 2014. Tube growth patterns and microbial mat-related lifestyles in the Ediacaran fossil *Cloudina*, Gaojiashan Lagerstätte, South China. *Gondwana Research* 25, 1008–1018.

Caron, J.-B., Scheltema, A., Schander, C. & Rudkin, D. 2006. A soft-bodied mollusk with radula from the Middle Cambrian Burgess Shale. *Nature* 442, 159–163.

Chen, L., Xiao, S., Pang, K., Zhou, C. & Yuan, X. 2014. Cell differentiation and germ–soma separation in Ediacaran animal embryo-like fossils. *Nature* 516, 238–241.

Conway Morris, S. 1998. *The Crucible of Creation*. Oxford University Press.

Dornbos, S.Q. 2006. Evolutionary palaeoecology of early epifaunal echinoderms: Response to increasing bioturbation levels during the Cambrian radiation. *Palaeogeography, Palaeoclimatology, Palaeoecology* 237, 225–239.

Dornbos, S.Q. & Chen, J.-Y. 2008. Community palaeoecology of the early Cambrian Maotianshan Shale biota: Ecological dominance of priapulid worms. *Palaeogeography, Palaeoclimatology, Palaeoecology* 258, 200–212.

Dornbos, S.Q., Bottjer, D.J. & Chen, J.-Y. 2005. Paleoecology of benthic metazoans in the Early Cambrian Maotianshan Shale biota and the Middle Cambrian Burgess Shale biota: evidence for the Cambrian substrate revolution. *Palaeogeography, Palaeoclimatology, Palaeoecology* 220, 47–67.

Droser, M.L. & Bottjer, D.J. 1993. Trends and patterns of Phanerozoic ichnofabrics. *Annual Reviews of Earth and Planetary Sciences* 21, 205–225.

Droser, M.L. & Gehling, J.G. 2008. Synchronous aggregate growth in an abundant new Ediacaran tubular organism. *Science* 319, 1660–1662.

Erwin, D.H., Laflamme, M., Tweedt, S.M., Sperling, E.A., Pisani, D. & Peterson, K.J. 2011. The Cambrian conundrum: Early divergence and later ecological success in the early history of animals. *Science* 334, 1091–1097.

Gehling, J.G. & Droser, M.L. 2012. Ediacaran stratigraphy and the biota of the Adelaide Geosyncline, South Australia. *Episodes* 35, 236–246.

Lefebvre, B. 2003. Functional morphology of stylophoran echinoderms. *Palaeontology* 46, 511–555.

Mata, S.A. & Bottjer, D.J. 2012. Facies control on Lower Cambrian wrinkle structure development and paleoenvironmental distribution, southern Great Basin, United States. *Facies*, DOI 10.1007/s10347-012-0331-3.

Noffke, N., Beukes, N., Bower, D., Hazen, M. & Swift, J.P. 2008. An actualistic perspective into Archean worlds – (cyano-)bacterially induced sedimentary structures in the siliciclastic Nhlazatse Section, 2.9 GA Pongola Supergroup, South Africa. *Geobiology* 5, pp. 5–20.

Payne, J.L, Boyer, A.G., Brown, J.H., Finnegan, S., Kowalewski, M., Krause, R.A., Jr.,, Lyons, S.K., McClain, C.R., McShea, D.W., Novack-Gottshall, P.M., Smith, F.A., Stempien, J.A. & Wang, S.C. 2009. Two-phase increase in the maximum size of life over 3.5 billion years reflects biological innovation and environmental opportunity. *Proceedings of the National Academy of Sciences* 106, 24–27.

Retallack, G.J. 2013. Ediacaran life on land. *Nature* 494, 89–92.

Seilacher, A. 1999. Biomat-related lifestyles in the Precambrian. *Palaios* 14, 86–93.

Tarhan, L.G. & Droser, M.L. 2014. Widespread delayed mixing in early to middle Cambrian marine shelfal settings. *Palaeogeography, Palaeoclimatology, Palaeoecology* 399, 310–322.

Xiao, S., Droser, M., Gehling, J.G., Hughes, I.V., Wan, B., Chen, Z. & Yuan, X. 2013. Affirming life aquatic for the Ediacara biota in China and Australia. *Geology* 41, 1095–1098.

Yin, A., Zhu, M., Davidson, E.H., Bottjer, D.J., Zhao, F. & Tafforeau, P. 2015. Sponge grade body fossil with cellular resolution dating 60 Myr before the Cambrian. *Proceedings of the National Academy of Sciences* 112, E1453–E1460.

Zimmer, C. 2006. *Evolution: The Triumph of an Idea*. Harper-Collins.

Additional reading

Erwin, D.H. & Valentine, J.W. 2013. *The Cambrian Explosion*. Roberts and Company.

Fedonkin, M.A., Gehling, J.G., Grey, K., Narbonne, G.M. & Vickers-Rich, P. 2007. *The Rise of Animals: Evolution and Diversification of the Kingdom Animalia*. Johns Hopkins University Press.

Knoll, A.H. 2003. *Life on a Young Planet*. Princeton University Press.

8 Phanerozoic level-bottom marine environments

Introduction

Ancient epicontinental sea and continental margin level-bottom environments from the nearshore and shallow subtidal to deeper-water settings represent most of the benthic settings preserved in Phanerozoic marine sedimentary rocks. They are commonly richly fossiliferous so that a variety of sources of data is available from which to understand the paleoecology and evolution of organisms that have lived there. Thus, from the fossil and stratigraphic record, these environments represent the most abundant source of information on the evolution of life and life's ecology. Understanding paleoecological conditions in these environments has also been very important as keys to the discovery of many natural resources that they preferentially contain and that are necessary for modern civilization.

Data collection and analysis

One of the key questions for Phanerozoic level-bottom environments is whether the body fossils that occur in strata deposited in these settings are preserved with enough integrity that one can actually reconstitute ancient communities and therefore understand how they evolved. A very

important work done by Kidwell (2001) has shown that indeed living assemblages are relatively well preserved as fossil assemblages from shoreline and shelf level-bottom settings. As shown in Fig. 8.1, these conclusions were drawn from analyses that were done from modern environments where surface and underlying sediment samples such as cores were taken, together with their living and fossil biota. This assessment was then made comparing living seafloor communities to subfossil assemblages preserved directly beneath them in the sediment column. This maxim holds true particularly for sampled shells greater than 1 mm in size, which is the typical size range encountered in paleocommunity studies. It is not true for samples with smaller shells because shells smaller than 1 mm in size tend to be larval. As larval presence in communities is pulsed and most larvae die before becoming adults, the timing of sampling living communities that recovers shells smaller than 1 mm can strongly affect the species composition in the resulting data sets.

Studies of ancient communities are most commonly done from these level-bottom continental margin environments. To collect a representative sample of an ancient community, a typical approach for these types of settings is to remove from an outcrop or core a bulk sample of rock or unlithified sediment with contained fossils. Bulk samples are made of a predetermined volume of rock or

Paleoecology: Past, Present and Future, First Edition. David J. Bottjer.
© 2016 John Wiley & Sons, Ltd. Published 2016 by John Wiley & Sons, Ltd.

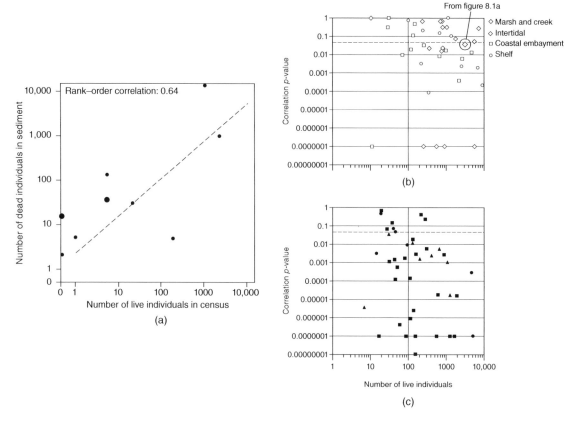

Figure 8.1 The fidelity of the fossil record for understanding ancient level-bottom communities. Kidwell assessed whether species composition and abundance for living and subfossil assemblages from the same location are similar. This included a comparison of life assemblages versus death assemblages for 85 collections representing a variety of marine settings. (a) Comparison between live abundances and dead abundances of 11 species at a tidal creek locality in California. Small points represent one species each; larger points each represent two species with the same live and dead abundances. The dashed line shows the trend in the relationship between live and dead abundances for species found in both the live and dead samples. Species that are more abundant in the live sample tend to be more abundant in the dead sample as well. The correlation between live and dead abundances, using a rank-order coefficient, is statistically significant with a p-value of 0.04

(p-values of 0.05 or less are considered to be statistically significant). Parts (b) and (c) illustrate live abundances and p-values for 85 comparisons like that of part (a), the horizontal dashed line marks the p-value of 0.05. (b) Forty-three collections made with sieve sizes of 1 mm or less. The comparison in part (a) is indicated by the large circled diamond, which is statistically significant. However, many of these comparisons do not produce significant correlations between live and dead abundances. (c) Forty-two collections made with sieve sizes greater than 1 mm. These greater sieve mesh sizes show most samples as statistically significant. It is likely that inclusion of smaller shells in samples, caused by such processes as an infusion of larvae into the population, skews results because in general smaller shells are typically more susceptible to postmortem destruction and death. From Foote and Miller (2006). Reproduced with permission from W.H. Freeman.

sediment from stratigraphic intervals of varying relatively small (10–15 cm) thicknesses that do not show evidence of any significant transport. From these bulk samples, the fossils which they contain are identified. The thinner the stratigraphic thickness of the bulk sample, the less time averaging will be involved (see Chapter 4). The volume of bulk samples can be determined through rarefaction analysis (Hammer and Harper 2006), where the optimal bulk sample size is when increasing the size yields few to no new species in the sample. Faunas of these bulk samples are equivalent to the subfossil samples that Kidwell (2001) has studied from modern level-bottom environments and therefore can be treated typically as relatively coherent samples of ancient communities.

There are a variety of quantitative approaches that are available to analyze samples of fossil assemblages (e.g., Hammer and Harper 2006). One of the easiest and most straightforward to use is the technique known as cluster analysis. This mathematical approach, although not statistical, joins assemblages from fossil samples that have the greatest similarity into groups (clusters) and can be used for any type of fossil sample. Cluster analysis of bulk samples which have undergone little postmortem transport can then undergo further ecological reconstruction in paleocommunity analysis. In addition, multivariate statistical approaches for defining and understanding the characteristics of ancient communities, including principal component analysis (PCA) and detrended correspondence analysis (DCA), are also widely utilized (e.g., Hammer and Harper 2006). A growing trend in paleocommunity studies has been to not only use presence–absence data but also abundance data (e.g., Clapham et al. 2006). Along with collecting your own original data, analysis of ancient communities can also be completed with data located in the Paleobiology Database (e.g., Bonuso and Bottjer 2008).

Certain transportation processes produce assemblages that preserve components of their original

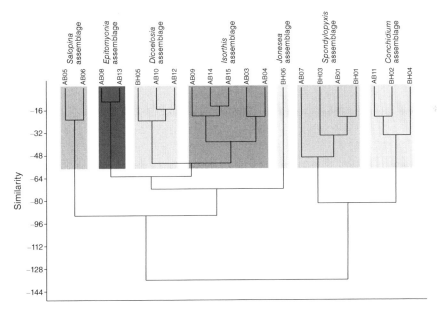

Figure 8.2 Cluster analysis of 20 brachiopod fossil samples (14,445 brachiopod specimens) from the Silurian Cape Philips Formation, Baillie Hamilton Island, and Cornwallis Island, Arctic Canada. Together with the known paleobathymetric distribution of these brachiopods from other regions, these transported brachiopod assemblages were then used to reconstruct sea-level changes during deposition of this formation. From Chen et al. (2012). Reproduced with permission from Elsevier.

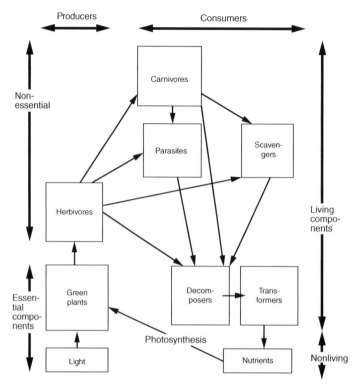

Figure 8.3 Schematic of a typical food web based on photosynthesis. Arrows show flow of materials through the food web. Green plants fix energy from light into organic molecules, and herbivores consume green plants. Carnivores consume herbivores, parasites and scavengers consume energy from herbivores and carnivores, and decomposers and transformers return nutrients to the environment. From Prothero (2003).

community composition. For example, Fig. 8.2 shows 20 Silurian fossil samples which are grouped by similarity via cluster analysis into six assemblages. These 20 samples came from debris and turbidity flows that had traveled from shallow water into a basin in which black shale was deposited. Other studies of Silurian brachiopod assemblages have led to a good understanding of the paleobathymetric distribution of Silurian brachiopods. Combined with this information, these assemblages were further analyzed to create a sea-level history for the studied formation. Assemblages with larger percentages of shallow-water brachiopods are interpreted to indicate a shallower part of the sea-level history, and assemblages with a greater percentage of deeper-water brachiopods indicate deeper parts of the sea-level trend.

Given the degree of preservation of the original community and the extent of data collection on the original community, trophic analysis of ancient communities can be a very useful approach. One can determine from studies of functional morphology the feeding relationships of the organisms within samples that have been collected. Thus, a food web can be constructed, as is portrayed in the theoretical example of Fig. 8.3, ranging from producers (plants) to herbivores to carnivores. Such an analysis is shown in more detail for an Ordovician paleocommunity assemblage depicted in Fig. 8.4, which illustrates the various component organisms within a trophic web with the top predator being the large nautiloid cephalopod. Extensive modeling approaches for trophic analysis are also available (e.g., Roopnarine 2009).

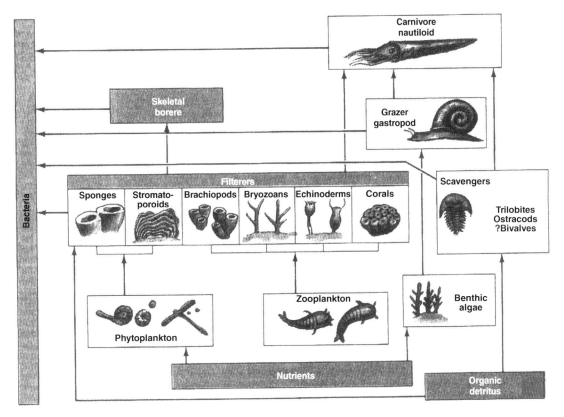

Figure 8.4 Reconstruction of an Ordovician food web. Like modern marine food webs, plants, and planktonic animals are the food for a variety of filter feeders (although those shown here are mostly extinct Paleozoic groups). Unlike modern ecosystems, there are relatively few deposit feeders (mainly trilobites) and very few predators (mainly straight-shelled cephalopods). The multiple levels of crustacean and fish predators that now characterize the ocean were a later development. From Prothero (2003), where data sources are indicated.

Nearshore and shelf-depth environments

A very significant source of paleontological information in level-bottom environments is provided by trace fossils. Much of the work on trace fossils that is done with identifying ichnofacies is from such environments. Ichnofabric analyses are also commonly produced from these types of ancient settings. For example, ichnofabric of nearshore sands is very intriguing to study. Nearshore sands usually encompass the Skolithos ichnofacies. As shown in Fig. 8.5, ichnograms from various nearshore sandstones of varying ages show the distribution in extent of bioturbation over geological time. Post-Paleozoic nearshore sandstones have a distribution of ichnofabric indices showing less overall bioturbation, as compared with the early Paleozoic examples, which contain packed beds of the vertical cylindrical burrow *Skolithos*, termed piperock, and have ichnofabric indices of fours and fives.

These variations in sedimentary fabric that can be determined and measured using ichnofabric indices from nearshore sandstones are illustrated schematically in Fig. 8.6. Here is shown the typical

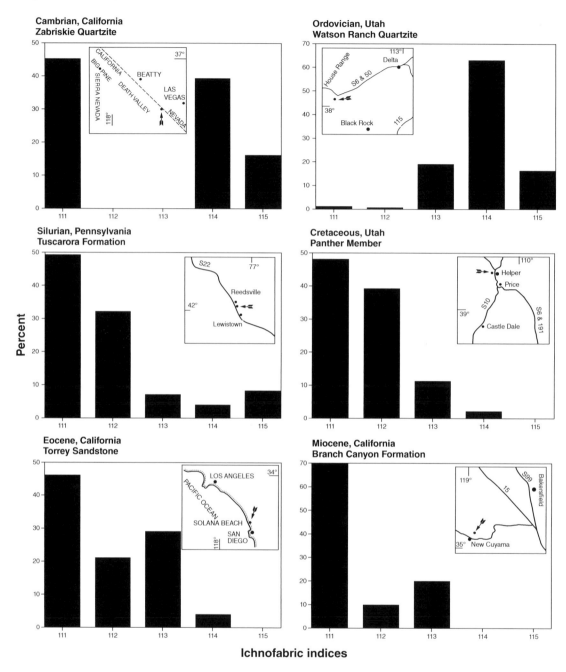

Figure 8.5 Distribution of ichnofabric indices from six Cambrian- to Miocene-aged high-energy nearshore sandstones for 10 m of measured section. The Cambrian Zabriskie Quartzite and the Ordovician Watson Ranch Quartzite have a relatively high percentage of beds with ii4 and ii5 because of the prevalence of *Skolithos* piperock at these sites. The Silurian Tuscarora Formation lacks piperock and hence thoroughly bioturbated beds. The Cretaceous Panther Member, Eocene Torrey Sandstone, and Miocene Branch Canyon Formation are characterized by the trace fossil *Ophiomorpha*, which rarely is involved in complete bioturbation of a bed. From Droser and Bottjer (1989). Reproduced with permission from the SEPM Society for Sedimentary Geology.

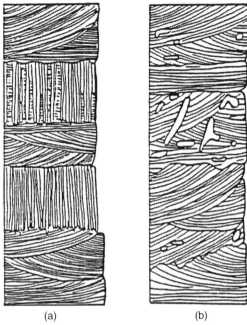

Figure 8.6 Schematic of characteristic distribution of Paleozoic and post-Paleozoic ichnofabrics within high-energy nearshore sandstones, based on data portrayed in Fig. 8.5. (a) Strata with *Skolithos* as the dominant trace fossil, commonly forming piperock, typical of many early Paleozoic examples. (b) Strata with *Ophiomorpha* as the dominant trace fossil, typical of many post-Paleozoic examples. From Droser and Bottjer (1989). Reproduced with permission from the SEPM Society for Sedimentary Geology.

fabric distribution in many early Paleozoic sedimentary rocks where nearshore sands contain piperock. In contrast, ichnofabric distribution in the post-Paleozoic, when *Skolithos* is a much rarer trace fossil and after the branching trace fossil *Ophiomorpha* had evolved and become common in such sandy environments, shows relatively lower amounts of bioturbation. This analysis shows how evolution has changed the distribution of sedimentary processes in nearshore sands due to the different kinds of animals that have been bioturbating in these environments.

Moving into deeper subtidal environments, there have been many studies done of the body

fossils that can be collected from these types of settings through the Phanerozoic. An elegant study by Robert Stanton and J. Robert Dodd is from the Pliocene of the Central Valley of California. Figure 8.7 shows the distribution of different fossils in the different community types that are found in these Pliocene settings. These community types have been determined through extensive sampling procedures followed by cluster analysis. Figure 8.8 shows the linkage of the distribution of fossil assemblages between different environments within these Pliocene deposits. This analysis (Figs. 8.7 and 8.8) gives a clear and robust understanding of the benthic invertebrate communities that lived in this Pliocene seaway.

Within level-bottom environments, sedimentary hardgrounds as well as rocks, shells, and wood have hosted encrusting as well as boring organisms. This is where the trace fossil record of boring organisms intersects with the body fossil record of encrusting organisms. Because encrusting organisms retain many of their original spatial relationships, they commonly offer unique opportunities for paleoecological reconstruction. Many encrusters commonly have mineralized skeletons, but many do not, so that taphonomic analyses of these ancient communities are also essential (Taylor and Wilson 2003).

Low-oxygen environments

Further traversing across shelf-depth environments, typically the progression down the slope and into deeper settings produces conditions where there is less dissolved oxygen in the water column. In Southern California, the abundance of organisms in Santa Monica Bay decreases with greater and greater depths, which has been linked to decrease in seawater oxygen content with depth (Fig. 8.9). An analysis of the widths of burrows from surface sediment box cores from shallow to deep water into various Southern California offshore basins also shows a decrease in size of burrows and thus burrowing organisms as water depths increase (Fig. 8.10). This trend has also been attributed to a

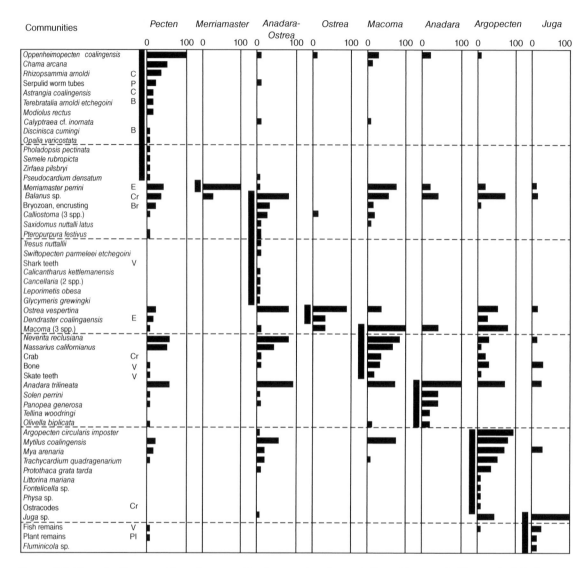

Figure 8.7 Macrofossil composition of communities in the Pliocene *Pecten* Zone from the Kettleman Hills of west-central California. Communities are recognized based on cluster analysis of the faunal data. Bar length is the percent of localities in each community at which each species is present. Species are arranged in decreasing order of frequency within a community. Species grouped together as indicated by a vertical bar to the left of a column are more common in that community than in the other communities. Letters to the right of the names identify nonmolluscan taxa: C, coral; P, polychaete; B, brachiopod, E: echinoderm; Cr, crustacean; V, vertebrate; Pl, plant. From Stanton and Dodd (1997). Reproduced with permission from Taylor and Francis.

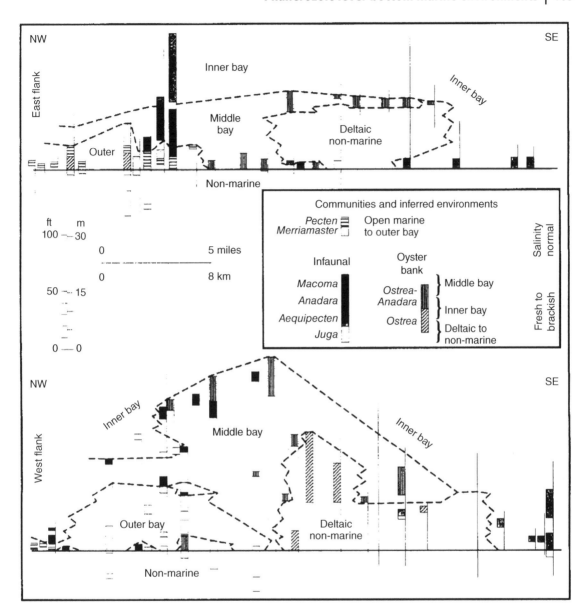

Figure 8.8 Distribution of communities (composition shown in Fig. 8.7) and paleoenvironmental reconstruction in the *Pecten* Zone along the east (top) and west (bottom) flanks of the Kettleman Hills of west-central California. The datum line is the base of the *Pecten* Zone. The vertical lines indicate the extent of stratigraphic section examined. From Stanton and Dodd (1997). Reproduced with permission from Taylor and Francis.

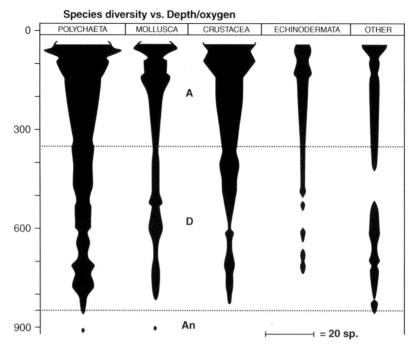

Figure 8.9 Trends in species richness from shallow to basinal depths in Santa Monica Basin, from offshore Southern California. Data was derived from faunal analysis of the upper portions of box core sediments, sieved with a 1 mm mesh screen. Depth on left axis is in meters; shelf break is at 150–200 m. The oxygen content above the upper dashed line is above 1.0 mL/L, representing aerobic environments. The oxygen content below this upper dashed line is between 0.1 and 1.0 mL/L, representing dysaerobic environments. Near elimination of macrobenthic elements at approximately 850 m is at the dysaerobic–anaerobic boundary in this basin, where oxygen content drops below 0.1 mL/L. A, aerobic; D, dysaerobic; An, anaerobic. From Savrda et al. (1984). Reproduced with permission from the American Association of Petroleum Geologists.

decrease in oxygen content with depth of the water overlying the seafloor.

These sorts of studies have been utilized to understand paleo-oxygen levels during deposition of ancient sedimentary rocks that were deposited in low-oxygen environments. A variety of biofacies schemes (e.g., Figs. 8.11 and 8.14) have been developed for understanding how to reconstruct ancient seafloor oxygen levels from benthic body fossils, as well as trace fossils, associated ichnofabric, and other sedimentary structures such as laminations. Figure 8.11, based on studies of modern and ancient low-oxygen environments, illustrates one such approach toward dividing various benthic environments into decreasing oxygen concentration, defining five biofacies including the unusual exaerobic biofacies.

Work done with Charles Savrda led to recognition of the exaerobic biofacies, which occurs where particular organisms have evolved to live on the margins of the anaerobic zone. Sometimes, these organisms are chemosymbiotic, respiring from oxygen available outside the anaerobic zone but utilizing methane or sulfide available in anaerobic environments for the chemosymbiotic bacteria which they host. An example of deposition of a modern exaerobic biofacies is found in Santa Barbara Basin off Southern California, where a

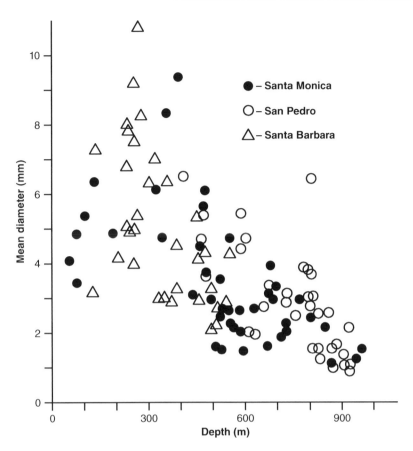

Figure 8.10 Decrease in burrow diameter with decrease in oxygenation of bottom water. Average burrow diameter was measured from X-radiographs of vertical box core slabs, with 6–10 measurements from each X-radiograph. The plot illustrates average burrow diameter and depth and therefore corresponding degree of oxygenation in Santa Monica, San Pedro, and Santa Barbara Basins, from offshore Southern California. Degree of oxygenation with depth for Santa Monica Basin is shown in Fig. 8.9, and that for San Pedro and Santa Barbara Basins shows the same trend but at slightly different depths. From Savrda et al. (1984). Reproduced with permission from the American Association of Petroleum Geologists.

ring of chemosymbiotic lucinid bivalves lives in the basin around the edge of the anaerobic zone. In the fossil record, a particularly illuminating study of the exaerobic biofacies was done for the Cambrian of the House Range in Utah. In Fig. 8.12 is shown a stratigraphic analysis where abundance of the trilobite *Elrathia* and average ichnofabric index are plotted against a stratigraphic column. This analysis shows that the mobile *Elrathia* was a common inhabitant of environments just outside the anaerobic zone, and thus, this association is a good example of an exaerobic biofacies.

In the past, many epicontinental seas had deeper basins, which were places where the overlying water column was stratified and water with low-oxygen content lay over the seafloor. One of the famous examples of this sort of ancient setting is the Posidonia Shale from the Jurassic of Germany. The Posidonia Shale is studied not only for its intriguing paleoecology but also because it represents a time

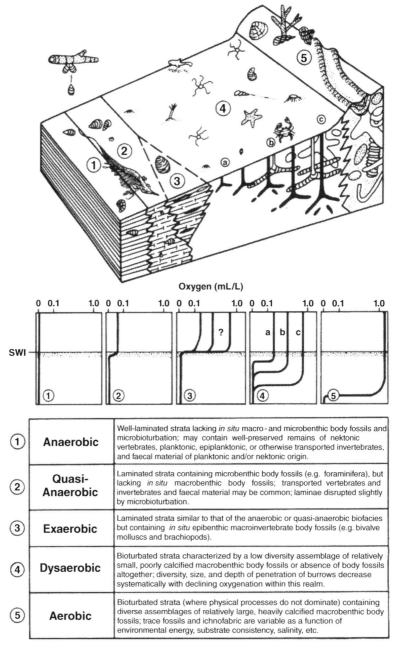

①	**Anaerobic**	Well-laminated strata lacking *in situ* macro- and microbenthic body fossils and microbioturbation; may contain well-preserved remains of nektonic vertebrates, planktonic, epiplanktonic, or otherwise transported invertebrates, and faecal material of planktonic and/or nektonic origin.
②	**Quasi-Anaerobic**	Laminated strata containing microbenthic body fossils (e.g. foraminifera), but lacking *in situ* macrobenthic body fossils; transported vertebrates and invertebrates and faecal material may be common; laminae disrupted slightly by microbioturbation.
③	**Exaerobic**	Laminated strata similar to that of the anaerobic or quasi-anaerobic biofacies but containing *in situ* epibenthic macroinvertebrate body fossils (e.g. bivalve molluscs and brachiopods).
④	**Dysaerobic**	Bioturbated strata characterized by a low diversity assemblage of relatively small, poorly calcified macrobenthic body fossils or absence of body fossils altogether; diversity, size, and depth of penetration of burrows decrease systematically with declining oxygenation within this realm.
⑤	**Aerobic**	Bioturbated strata (where physical processes do not dominate) containing diverse assemblages of relatively large, heavily calcified macrobenthic body fossils; trace fossils and ichnofabric are variable as a function of environmental energy, substrate consistency, salinity, etc.

Figure 8.11 Determination of oxygenation levels in ancient environments from analysis of sedimentary fabrics, trace fossils, and body fossils, synthesized from a variety of modern and ancient settings. This approach is summarized as characteristics of five oxygen-related biofacies, shown here with lateral oxygen gradients along the seafloor (top) and vertical oxygen gradients across the sediment–water interface (SWI) (middle). Schematic oxygen profiles (at STP) are intended to reflect the relationships between bottom-water oxygenation and the time-averaged position of the redox potential discontinuity (the boundary below which dissolved oxygen does not exist). From Bottjer and Savrda (1993). Reproduced with permission from John Wiley & Sons.

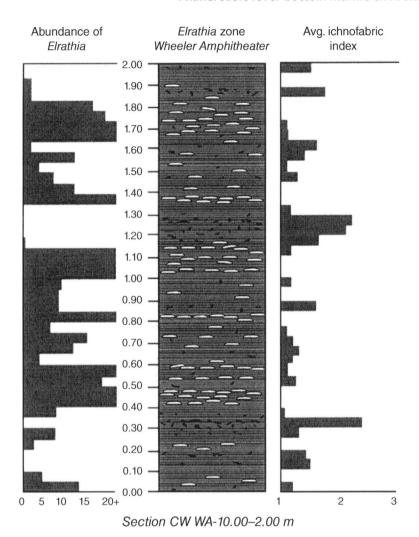

Section CW WA-10.00–2.00 m

Figure 8.12 Stratigraphic section, 2 m thick, from the Middle Cambrian Wheeler Formation (Wheeler Amphitheater, House Range, Utah, United States), with abundance of the trilobite *Elrathia kingii* (normalized to 50 cm^2) and average ichnofabric index. In this section, 440 beds were examined; schematic data shown represent bins spanning 5 cm of vertical section. *E. kingii* occur in maximum abundance in unbioturbated (ii1) and much less commonly in weakly bioturbated (ii2) beds, with no occurrence in beds with greater ichnofabric indices recorded. Thus, using 5 cm bins, the abundance of *E. kingii* and the ichnofabric index in this section are negatively correlated (correlation coefficient = −0.52). On the seafloor, exaerobic organisms live adjacent to anaerobic environments, the habitat most likely indicated by this stratigraphic distribution of *E. kingii* and ichnofabric. From Gaines and Droser (2003). Reproduced with permission from the Geological Society of America.

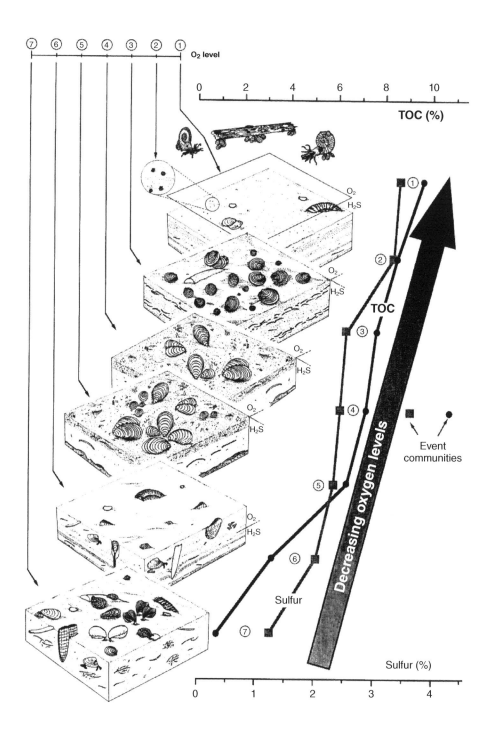

Figure 8.13 Oxygen-related community and geochemical analysis of the lower Toarcian (Jurassic) Posidonia Shale (southwest Germany). Oxygen availability was highly variable and ranged from long-term anoxic (O_2 level 1) to long-term oxic (O_2 level 7). The position of the redox boundary fluctuated, oxygen concentration changed, and the duration of oxygenated phases varied. Combined paleoecological and geochemical methods were used to reconstruct a relative paleo-oxygen curve. Average values of organic carbon and sulfur content display the same trend. Long phases without any benthic fauna (O_2 level 1: macrofauna is only represented by pseudoplanktonic and nektonic organisms) were interrupted by very short colonization events, indicated by tiny juvenile bivalves, which died soon after colonizing the sediment surface (O_2 level 2). Poikilo-oxic conditions allowed temporary benthic colonization of opportunistic species (*Bositra buchi*,

Meleagrinella substriata) probably during the winter months, when oxygenated bottom water prevailed (O_2 level 3). Monospecific colonization of *Pseudomytiloides dubius* took place during oxygenated periods lasting several months to years (O_2 level 4). Note that horizons with *Pseudomytiloides*-event communities show significantly higher TOC and sulfur content. The *Pseudomytiloides dubius/Discina papyracea* association indicates oxygen availability of several years (O_2 level 5). Thin bioturbated horizons with a small benthic infauna point to a redox boundary that was situated below the sediment–water interface for several years (O_2 level 6). Long-term stable conditions with well-oxygenated bottom water favored a diverse benthic infauna and epifauna (*Rhynchonella amalthei/Plicatula spinosa* association; O_2 level 7). From Rohl et al. (2001). Reproduced with permission from Elsevier.

of global levels of anoxia in various geographic settings. Examination in detail of the benthic assemblages that occur in the Posidonia Shale and coupling those analyses with total organic content and sulfur percentage, as seen in Fig. 8.13, shows that the benthic communities trend from lower to higher oxygen concentrations during their deposition. Incorporating these analyses into a biofacies approach, Fig. 8.14 shows a schematic indicating a biofacies interpretation for Posidonia Shale depositional environments. Biofacies range from purely laminated sediments to partially bioturbated sediments, which contain some fossils, to greater amounts of bioturbation and fossil content, thus proceeding from anoxic to oxygenated environments in the Posidonia Shale. These benthic settings can then be related also to the overlying water column and the organisms which inhabited it, including fish, ichthyosaurs, and ammonites. As discussed in Chapter 4, this type of setting provides the conditions for the development of a lagerstätten, and the Posidonia Shale is well-known as a major Jurassic lagerstätten.

In low-oxygen environments, as sediment accumulates, tiered trace fossils produce a variety of intriguing ichnofabrics. If there is stable oxygenation, the ichnofabric will be one in which the deeper-tiered trace fossils crosscut the shallower-tiered trace fossils. Work done with Charles Savrda shows that these various tiering patterns of overlapping relationships in sedimentary rocks deposited under low but variable oxygen concentrations can be reconstructed as seen in Fig. 8.15. This schematic depicts the sorts of relationships of overlapping trace fossils that one might find with stable oxygenation, gradual deoxygenation, gradual oxygenation, and rapid changes in oxygenation. These various overlapping crosscutting relationships can then be analyzed together with measurements on burrow width to produce an oxygenation curve for different sedimentary environments from the ichnofabric and trace fossil record. For many stratigraphic settings, such as for the Miocene Monterey Formation of California or the Niobrara Formation of Colorado (Fig. 8.16), this provides a relatively high-resolution determination of how oxygen fluctuated on the seafloor in these ancient environments.

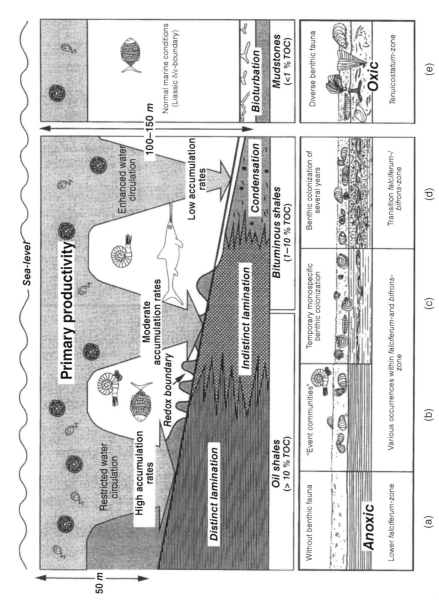

Figure 8.14 Biofacies analysis of the Jurassic Posidonia Shale. This schematic shows a sequence of biofacies during deposition of the lower Toarcian Posidonia Shale (southwest Germany) related to water depth and redox conditions. These biofacies incorporate analyses shown in Fig. 8.13 and are similar to but defined slightly differently from those shown in Fig. 8.11. (a) Restricted water circulation during sea-level lowstand led to long-term anoxic conditions excluding benthic faunas, producing distinctly laminated sediments with high organic carbon content. (b) Anoxia was intermittently interrupted by very short oxygenated periods, indicated by event communities. (d) Benthic colonization of several years occurred during high sea level with enhanced water circulation; sediments are characterized by condensation and reduced TOC values. (c) An intermediate position of sea level allowed temporary benthic colonization, and indistinct laminated sediments were deposited. (e) Bioturbated mudstones with low organic carbon content and a diverse benthic fauna point to normal marine aerobic conditions during the *tenuicostatum* zone. From Rohl et al. (2001). Reproduced with permission from Elsevier.

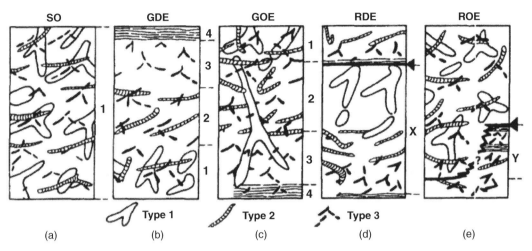

Figure 8.15 Trace fossil model for reconstructing paleo-oxygenation in ancient bottom waters. These schematic diagrams illustrate tiering relationships expected in an initially stable, well-oxygenated (SO) sediment column (a); tiering relationships predicted if degree of oxygenation decreases (b) or increases (c) gradually through time (GDE and GOE); and tiering relationships predicted if degree of oxygenation decreases (d) or increases (e) abruptly (RDE and ROE). Numbered intervals between dashed lines to the right of columns a–c delineate oxygen-related ichnocoenosis (ORI) units, boundaries of which represent threshold levels at which certain trace types appear or disappear. In d, X delineates "frozen tiered profile," which occurs after a rapid oxygenation decrease. In e, Y delineates interval for which record may be destroyed by subsequent, penetrative bioturbation after a rapid oxygenation increase. Types 1–3 are analogous to *Thalassinoides*, *Zoophycos*, and *Chondrites*. An ichnocoenosis is a trace fossil assemblage. From Savrda and Bottjer (1986). Reproduced with permission from the Geological Society of America.

Summary

With the availability of this broad array of evidence, level-bottom marine ecosystems are some of the best understood in Earth history. Studies of these environments have thus provided significant insights on how benthic ecology has evolved. Yet much work remains to be done, particularly on the sedimentary context of these fossil assemblages. For example, continued study of pelagic sediments is leading to a more in-depth appreciation on the meaning of trace fossils and ichnofabrics in these settings (e.g., Savrda 2014). Similarly, the new field of stratigraphic paleobiology emphasizes the effects of stratigraphic architecture upon our interpretation of fossil assemblages from these environments (Patzkowsky and Holland 2012). Integration of geochemical approaches with fossil data has also become increasingly valuable (e.g., Boyer et al. 2011) to understanding the environmental setting of these ancient ecosystems.

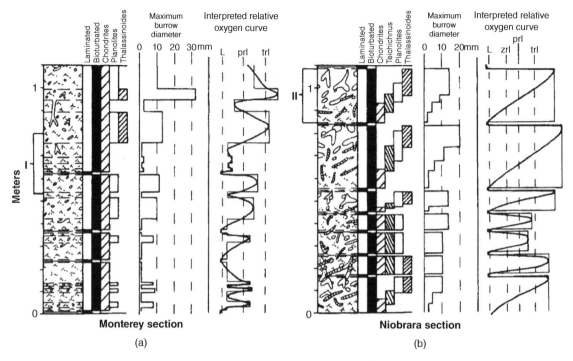

Figure 8.16 Utilization of trace fossil model for interpreting paleo-oxygenation in ancient depositional environments to produce relative oxygen curves. These interpreted relative oxygenation curves were derived from high-resolution vertical-sequence analyses of general sediment fabric, trace fossil composition, crosscutting relationships, and size parameters in sections of the Monterey Formation at Newport Lagoon, California (a) and the Niobrara Formation (top of Fort Hays Member) at Lyons, Colorado (b). *Teichichnus* column in b represents both *Teichichnus* and *Zoophycos*. Ichnologic approach defined in Fig. 8.15 has led to delineation of ORI unit boundaries, indicated here by dashed lines in graphic columns. Several reference lines are employed in the construction of the relative oxygenation curves. L, laminated reference line; prl, *Planolites* reference line; trl, *Thalassinoides* reference line; and zrl, *Teichichnus/Zoophycos* reference line. Although qualitative and interpretive, curves accurately reflect differences in relative degree of oxygenation and rates and magnitudes of change. Interval I in a and Interval II in b are illustrated in Savrda and Bottjer (1986). From Savrda and Bottjer (1986). Reproduced with permission from the Geological Society of America.

References

Bonuso, N. & Bottjer, D.J. 2008. A test of biogeographical, environmental, and ecological effect on Middle and Late Triassic brachiopod and bivalve abundance patterns. *Palaios* 23, 43–54.

Bottjer, D.J. & Savrda, C.E. 1993. Oxygen-related mudrock biofacies. *In* Wright, V.P. (ed.), *Sedimentology Review/1*, Blackwell Scientific Publications, pp. 92–102.

Boyer, D.L., Owens, J.D., Lyons, T.W. & Droser, M.L. 2011. Joining forces: Combined biological and geochemical proxies reveal a complex but refined high-resolution palaeo-oxygen history in Devonian epeiric seas. *Palaeogeography, Palaeoclimatology, Palaeoecology* 306, 134–146.

Chen, P., Jin, J. & Lenz, A.C. 2012. Palaeoecology of transported brachiopod assemblages embedded in black shale, Cape Phillips Formation (Silurian), Arctic Canada. *Palaeogeography, Palaeoclimatology, Palaeoecology* 367–368, 104–120.

Clapham, M.E., Bottjer, D.J., Powers, C.M., Bonuso, N., Fraiser, M.L., Marenco, P.J., Dornbos, S.Q. & Pruss, S.B. 2006. Assessing the ecological dominance of Phanerozoic marine invertebrates. *Palaios* 21, 431–441.

Droser, M.L. & Bottjer, D.J. 1989. Ichnofabric of sandstones deposited in high-energy nearshore environments: Measurement and utilization. *Palaios* 4, 598–604.

Foote, M. & Miller, A.I. 2006. *Principles of Paleontology, 3rd Edition*. W.H. Freeman and Company.

Gaines, R.R. & Droser, M.L. 2003. Paleoecology of the familiar trilobite *Elrathia kingii*: An early exaerobic zone inhabitant. *Geology* 31, 941–944.

Hammer, O. & Harper, D. 2006. *Paleontological Data Analysis*. Blackwell Publishing.

Kidwell, S.M. 2001. Preservation of species abundance in marine death assemblages. *Science* 294, 1091–1094.

Patzkowsky, M.E. & Holland, S.M. 2012. *Stratigraphic Paleobiology: Understanding the Distribution of Fossil Taxa in Time and Space*. The University of Chicago Press.

Prothero, D.R. 2003. *Bringing Fossils to Life*. McGraw Hill Higher Education.

Rohl, H.-J., Schmid-Rohl, A., Oschmann, W., Frimmel, A. & Schwark, L. 2001. The Posidonia Shale (Lower Toarcian) of SW-Germany: an oxygen-depleted ecosystem controlled by sea level and palaeoclimate. *Palaeogeography, Palaeoclimatology, Palaeoecology* 165, 27–52.

Roopnarine, P.D. 2009. Ecological modeling of paleocommunity food webs. *In* Dietl, G.P. & Flessa, K.W. (eds.), *Conservation Paleobiology: Using the Past to Manage for the Future*. The Paleontological Society, Boulder, *Papers Volume 15*. pp. 195–220.

Savrda, C.E. 2014. Limited ichnologic fidelity and temporal resolution in pelagic sediments: Paleoenvironmental and paleoecologic implications. *Palaios* 29, 210–217.

Savrda, C.E. & Bottjer, D.J. 1986. Trace-fossil model for reconstruction of paleo-oxygenation in bottom waters. *Geology* 14, 3–6.

Savrda, C.E., Bottjer, D.J. & Gorsline, D.S. 1984. Development of a comprehensive oxygen-deficient marine biofacies model: Evidence from Santa Monica, San Pedro, and Santa Barbara Basins, California Continental Borderland. *The American Association of Petroleum Geologists Bulletin* 68, 1179–1192.

Stanton, R.J. & Dodd, J.R. 1997. Lack of stasis in late Cenozoic marine faunas and communities, central California. *Lethaia* 30, 239–256.

Taylor, P.D. & Wilson, M.A. 2003. Palaeoecology and evolution of marine hard substrate communities. *Earth-Science Reviews* 62, 1–103.

Additional reading

Brenchley, P.J. & Harper, D.A.T. 1998. *Palaeoecology: Ecosystems, Environments, Evolution*. Chapman and Hall.

McKerrow, W.S. 1982. *The Ecology of Fossils: An Illustrated Guide*. The MIT Press.

Valentine, J.W. 1973. *Evolutionary Paleoecology of the Marine Biosphere*. Prentice Hall.

9 Reefs, shell beds, cold seeps, and hydrothermal vents

Introduction

The stratigraphic record is well known for accumulations of carbonate rocks that are composed of skeletons of invertebrate organisms as well as carbonate mud and cements. These accumulations in sedimentary sequences have apparent concentrations of carbonate organisms and also at times exhibit features that demonstrate an original topographic relief on the seafloor. As these stratigraphic features have high concentrations of fossils, they are commonly a subject of intense study by paleontologists and paleoecologists. And, as has been discussed in Chapter 3, paleoecological models to interpret their mode of formation have developed over time. Because of the content of large macroinvertebrate fossils, these stratigraphic features also have a significant amount of original porosity and permeability and can be the location of significant petroleum reserves. Thus, they are commonly well studied for their economic significance.

Reefs

Such accumulations interpreted as fossil reefs in the stratigraphic record are commonly also surrounded by carbonate rocks. Reef facies typically consist of a core that can include framework-building organisms which create a topographic high on the seafloor, with erosion and transport of these framework organisms producing flank deposits on the sides (Fig. 9.1a). Thus, overall stratigraphically, there is a pinch-out of beds of the surrounding lithology as they approach the reef deposit. As also shown in Fig. 9.1, reefs are common in shallow-water environments where photosynthesis can occur, including that for photosymbionts in such organisms as corals. As discussed in Chapter 3, reefs also occur in deepwater environments (Fig. 9.1) where, although photosynthesis is not possible, conditions lead to a concentrated accumulation of frame-building organisms also producing a topographic high on the seafloor. The development of reefs is controlled by organic growth, sedimentation from the aggradation of reef components and surrounding sediments, as well as cementation in these commonly carbonate-saturated seawater environments (Fig. 9.1b). Reef cavities that develop have their own faunal constituents and also can provide much of the porosity that leads to reefs being good petroleum reservoirs (Fig. 9.1).

There is a significant literature on reef concepts as understood from the stratigraphic record. Reef was originally a word used for a topographic high on the seafloor that was a hard rocky surface of actual outcrop or debris (Fig. 9.2). Stratigraphic reefs are

Paleoecology: Past, Present and Future, First Edition. David J. Bottjer.
© 2016 John Wiley & Sons, Ltd. Published 2016 by John Wiley & Sons, Ltd.

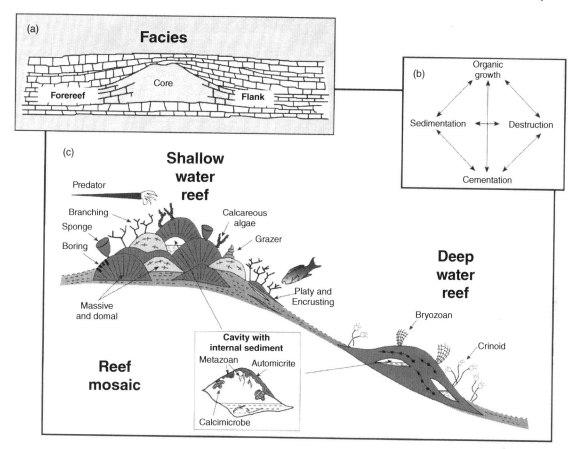

Figure 9.1 Development of reefs in modern and ancient environments. Schematic diagrams illustrating (a) cross-sectional geometry of a typical reef as exposed in outcrop, (b) complex relationships between biological and sedimentological processes that control reef composition, and (c) ecological and sedimentological attributes of shallow-water and deepwater reefs. From James and Wood (2010), where data sources are indicated. Reproduced with permission from the Geological Association of Canada. (*See insert for color representation.*)

typically found in carbonate rocks and are convex upward, commonly with characteristic fossils and skeletal debris that differs from the surrounding rocks. Mud mounds are also positive stratigraphic features that consist more commonly of carbonate mud with a reduced component of skeletal material, and these are found more likely in deeper water. Degraded reefs can also be identified by the presence of carbonate skeletons that would have formed in a reef (Fig. 9.2). Development of this reef classification has been driven by efforts to understand ancient reefs and whether they represent ecological reefs, which include such features as rigidity and the ability to resist waves, the presence of topographic relief and organic framework, a fairly high diversity of different ecological groups, and also attached organisms living in close association (Fig. 9.2).

One can further examine the concept of reefs based on reef fabric as shown in Fig. 9.3. Skeletal reefs are dominated by the skeletons of typical reef-building organisms (Fig. 9.3). There is a characteristic fabric for reefs that not only includes skeletal

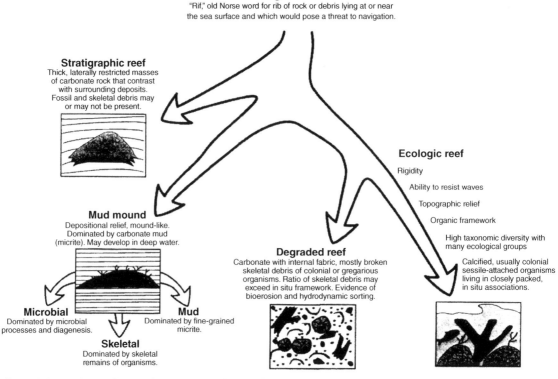

Figure 9.2 Variability of the reef concept in modern and ancient environments. From Stanley (2001). Reproduced with permission from Springer Verlag.

debris but also a significant presence of microbes and associated processes that form carbonate mud fabrics. Also included in reef classifications are those made of stromatolites and thrombolites as well as mud mounds, which are almost completely dominated by microbial processes.

The organisms that have built reefs have varied remarkably through time. As discussed in Chapters 6 and 7, stromatolites and stromatolite reefs form much of the most significant evidence for Precambrian life. Through much of the Precambrian, a great variety of stromatolite microbial reefs can be found (Figs. 7.1 and 9.4). These reefs have a fascinating variety of fabrics involving microbial fabrics as well as various cements. In Precambrian stromatolite reefs, there can also be observed a broad variety of stromatolite morphologies which are zoned across reef environments (Figs. 7.1 and 9.4).

The first metazoans that built reefs in the Phanerozoic are groups with carbonate skeletons that are allied with the sponges, including archaeocyaths (Fig. 9.5) and radiocyaths (Fig. 9.6). These lower Cambrian reefs have a lot of the features that are found commonly in modern-day reefs, including a variety of morphologies of archaeocyathids, typically the primary frame-building organisms for the reefs, as well as frame-building algae and cavities in which mobile and encrusting organisms lived (Fig. 9.6).

Later in the mid-Paleozoic are reefs built commonly by another group with a calcareous skeleton that is allied with the sponges, the stromatoporoids. A schematic of these stromatoporoid reefs is found in Fig. 9.7. Associated with the stromatoporoids are typical Paleozoic corals, such as rugose and tabulate corals, and many other kinds of organisms including crinoids and nautiloid cephalopods.

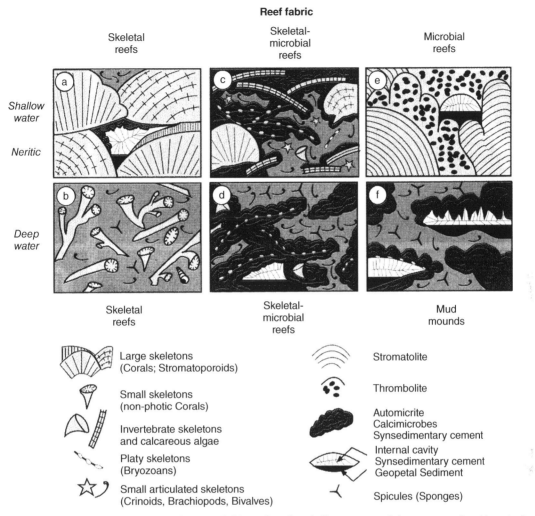

Reef fabric

Skeletal reefs | Skeletal-microbial reefs | Microbial reefs

Shallow water

Neritic

Deep water

Skeletal reefs | Skeletal-microbial reefs | Mud mounds

Large skeletons (Corals; Stromatoporoids)

Small skeletons (non-photic Corals)

Invertebrate skeletons and calcareous algae

Platy skeletons (Bryozoans)

Small articulated skeletons (Crinoids, Brachiopods, Bivalves)

Stromatolite

Thrombolite

Automicrite Calcimicrobes Synsedimentary cement

Internal cavity Synsedimentary cement Geopetal Sediment

Spicules (Sponges)

Figure 9.3 Schematic diagrams of the different rock fabrics found in shallow-water and deepwater reefs, with typical biotic and abiotic components. From James and Wood (2010), where data sources are indicated. Reproduced with permission from the Geological Association of Canada. (*See insert for color representation.*)

The end-Permian mass extinction saw the demise of tabulate and rugose corals and the elimination of metazoan reefs for the early part of the following Early Triassic (Figs. 13.6–13.9). However, by the Middle Triassic, scleractinian corals had evolved and began to become components of reefs. The Late Triassic is a time of extensive coral growth, and along with sponges, they built enormous reef deposits found in the Austrian Alps (Fig. 9.8). These reefs had lagoons with sponges and phaceloid coral gardens,

a reef crest with abundant sponges and cements, and a reef front with both corals and sponges (Fig. 9.9). Microbial fabrics are found throughout, most importantly in the back reef and lagoon.

Later, in the Mesozoic, sponges and corals became less dominant as reef builders in carbonate systems with the rise of the rudist bivalves. Huge accumulations of these bivalves, like that shown in Fig. 9.10, formed from the Jurassic to the end of the Cretaceous. Rudists had inequivalved shells with

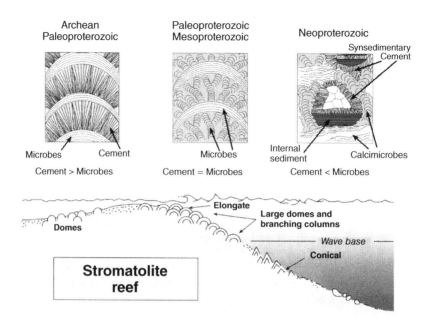

Figure 9.4 Precambrian stromatolite reefs. Schematic diagrams show Archean–Neoproterozoic reef fabrics of cement and microbial features (upper) and a zoned Proterozoic marginal reef (lower). In the Archean–Paleoproterozoic, the amount of synsedimentary cement was generally greater than the microbial content; in the later part of the Paleoproterozoic and Mesoproterozoic, they were about equal; in the Neoproterozoic, microbes and calcimicrobes were volumetrically more abundant than synsedimentary cement. Note characteristic trends in stromatolite morphology across reef environments. From James and Wood (2010), where data sources are indicated. Reproduced with permission from the Geological Association of Canada. (*See insert for color representation.*)

Figure 9.5 Branching archaeocyath sponge and surrounding sediment, including archaeocyath fragments. This image is from an outcrop of the lower Cambrian Poleta Formation in Esmeralda County, Nevada, USA. Branches of archaeocyath are ~5 mm wide. Additional information can be found in Rowland (1984). Photograph by Stephen M. Rowland. Reproduced with permission. (*See insert for color representation.*)

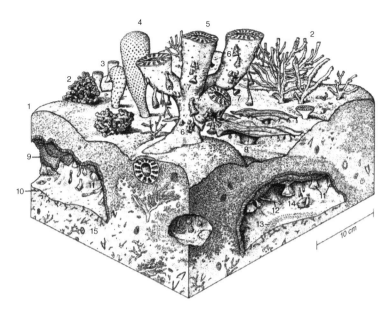

Figure 9.6 Reconstruction of a lower Cambrian reef. This reconstruction is of a reef community from Zuune Arts, western Mongolia (Atdabanian). (1) *Renalcis* (calcified cyanobacterium); (2) branching archaeocyath sponges; (3) solitary cup-shaped archaeocyath sponges; (4) chancelloriid; (5) radiocyaths; (6) small archaeocyath sponges; (7) "coralomorphs"; (8) *Okulitchicyathus* (archaeocyath sponge); (9) fibrous cement; (10) microburrows (traces of a deposit-feeder); (11) cryptic archaeocyaths and coralomorphs; (12) cribricyaths; (13) trilobite trackway; (14) cement botryoid; (15) sediment with skeletal debris (Copyright John Sibbick). From Wood (2000). Reproduced with permission from Oxford University Press.

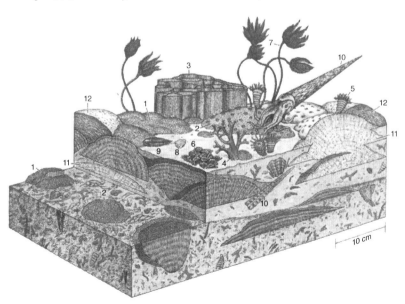

Figure 9.7 Reconstruction of a Silurian reef. This reef is composed mostly of stromatoporoids and tabulate corals. (1) Tabulate coral (*Favosites*); (2) tabulate coral (*Heliolites*); (3) tabulate coral (*Halysites*); (4) bryozoan; (5) rugose coral; (6) spiriferid brachiopod; (7) crinoid; (8) brachiopod; (9) trilobite; (10) orthocone nautiloid; (11) stromatoporoid; (12) thrombolite (Copyright John Sibbick). From James and Wood (2010), where data sources are indicated. Reproduced with permission from the Geological Association of Canada. (*See insert for color representation.*)

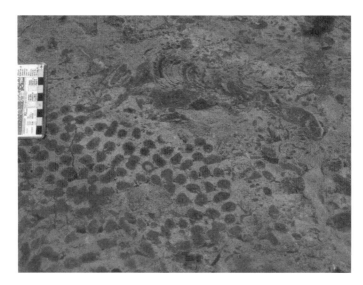

Figure 9.8 Cut slab of the upper Rhaetian (Late Triassic) Adnet reef from the Tropfbruch quarry (Oberrhätriffkalk Formation) near Adnet, Salzburg, Austria. Filling the lower left quadrant is a cross section through a colony of *Retiophyllia* sp. phaceloid corals (subcircular, dark gray spots). Visible as the largest fossil in the upper right hand quadrant is a sphinctozoan sponge. These are surrounded by reef sediment including solitary corals. Scale is in centimeters. Additional information can be found in Bernecker et al. (1999). Photograph by Rowan C. Martindale. Reproduced with permission. (*See insert for color representation.*)

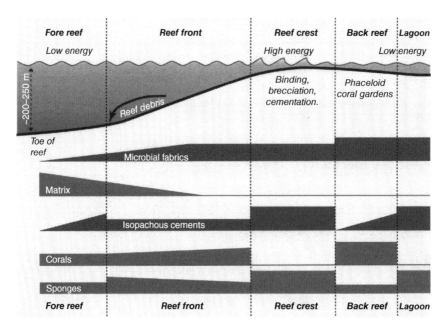

Figure 9.9 Compositional trends in an enormous Upper Triassic reef. This is an idealized environmental transect (top) 5 km long through the Rhaetian (Upper Triassic) Gosausee reef (Gosau, Austria), one of the Dachstein reefs, showing trends in microfacies composition for different reef facies (bottom). Microbial fabrics are common, and scleractinian corals and hypercalcified sponges are the primary metazoan components; phaceloid corals have multiple tubular corallites, each with a polyp, united at the base. From Martindale et al. (2013). Reproduced with permission from Elsevier. (*See insert for color representation.*)

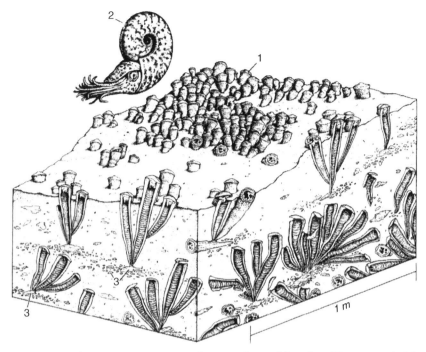

Figure 9.10 Reconstruction of an Upper Cretaceous rudist aggradation. (1) Rudists (*Vaccinites* sp.); (2) ammonite; (3) shell lags (Copyright John Sibbick). From Wood (2000). Reproduced with permission from Oxford University Press.

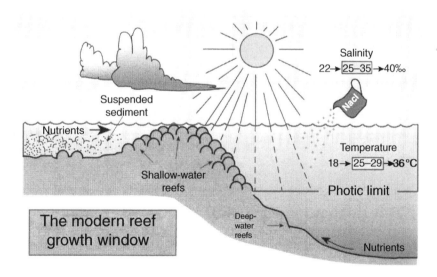

Figure 9.11 Schematic of the environmental parameters affecting the growth of modern coral reefs. Although reefs are typically low nutrient systems, nutrients can be supplied to shallow reefs from runoff, including suspended sediment, and to deepwater reefs through upwelling. Other important components include temperature, salinity, and the availability of sunlight. From James and Wood (2010). Reproduced with permission from the Geological Association of Canada. (*See insert for color representation.*)

Figure 9.12 Reconstruction of a modern reef from the Indo-Pacific. (1) Brain coral (*Leptoria phrygia*); (2) feather star (*Comanthus bennetti*); (3) parrotfish (*Scarus* sp.); (4) staghorn coral (*Acropora* sp.); (5) emperor angelfish (*Pomacanthus imperator*); (6) gorgonian; (7) vase sponge (*Callyspongia* sp.); (8) anemone with clown fish; (9) giant clam (*Tridacna gigas*); (10) encrusting corals (*Montipora* and *Hydnophora*); (11) brittle star (*Ophiarachella gorgonia*); (12) and (13) sea urchins; (14) cowrie; (15) sea cucumber (*Thelenota ananas*); (16) sea star; (17) boring bivalve (*Lithophaga*); (18) cement botryoids; (19) internal sediment; (20) cone shell (*Conus textile*); (21) wrasse (*Coris gaimard*) (Copyright John Sibbick). From Wood (2000). Reproduced with permission from the Oxford University Press.

Figure 9.13 Bedding plane view of a Middle Triassic shell bed composed primarily of the terebratulid brachiopod *Aulacothyroides liardensis*. This shell bed occurs within the Liard Formation outcropping along the shore of Williston Lake, British Columbia, Canada. The Liard Formation is a siliciclastic succession containing numerous shell beds, many like this one, which were deposited in a shallow shelf setting with a storm-dominated shoreline. Knife is 3.5 cm long. Additional information can be found in Greene et al. (2011). Photograph by David J. Bottjer. (*See insert for color representation.*)

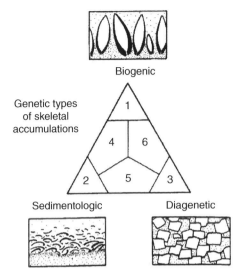

Biogenic

Genetic types
of skeletal
accumulations

1

4   6

2   5   3

Sedimentologic   Diagenetic

Figure 9.14 Processes that generate shell beds. This shows a conceptual framework for genesis of nonreef skeletal concentrations based on three end-member sets of concentrating processes. Biogenic concentrations (area 1) are produced by the gregarious behavior of skeletonized organisms (intrinsic biogenic) or by the actions of other organisms (extrinsic biogenic). Sedimentologic concentrations (area 2) form through hydraulic reworking of hardparts as particles and/or through nondeposition or selective removal of sedimentary matrix. Diagenetic concentrations (area 3) include residues of concentrated skeletal material along pressure solution seams and compaction-enhanced fossil horizons. Mixed origin concentrations (areas 4–6) reflect the strong influence of two or more different kinds of processes, for example, hydraulic overprinting of a biogenic precursor concentration. End-member concentrations (areas 1–3) can record single or multiple events of skeletal concentration; beds of mixed origin will most commonly reflect more than one episode of skeletal concentration. Concentrations of any of the six types can form rapidly (a few hours) or very slowly (hundreds to thousands of years); long-term concentrations will typically be mixed in origin. From Kidwell et al. (1986). Reproduced with permission from the SEPM Society for Sedimentary Geology.

a lower conical valve and an upper relatively flat valve. Many of them were semi-infaunal, employing a gregarious life habit (Fig. 9.10). The diverse rudists also include epifaunal and cementing forms.

Rudist bivalves became extinct at the end-Cretaceous mass extinction, and the Cenozoic is an era in which scleractinian corals once again are found to be dominant. Scleractinian corals and the reefs they build from Cenozoic deposits are typically interpreted using actualism from studies of modern reefs (Fig. 9.11). Modern shallow-water reefs grow typically with abundant sunlight and temperature conditions between 25 and 29 °C. They include a broad variety of corals as well as the photosymbiotic bivalve *Tridacna*. As is evident in Fig. 9.12, a notable component of boring and encrusting organisms has also existed in Cenozoic reefs.

Shell beds

Other types of skeletal accumulations in the fossil record do not form from preferential buildups on the seafloor. These skeletal accumulations are horizontal flat to lens-shaped features which commonly do not include reef-building or seep organisms. These have been termed shell beds (Fig. 9.13) and are a source of a great variety of paleobiological and paleoecological information. Figure 9.14 shows a characterization of the various types of processes that contribute to the development of such skeletal accumulations. These include nonreef biogenic accumulations, sedimentologic accumulations, and diagenetic accumulations of fossils. Shell beds are common in carbonate as well as siliciclastic sedimentary settings.

Modeling the different processes that create shell beds and how shell beds and these processes are distributed from shallow to deep water shows that different skeletal concentrations dominate in different environments. In tidal flat and lagoon environments, both biogenic and sedimentologic skeletal concentrations are common, with sedimentologic concentrations dominating in beach, shoal, and inner shelf environments (Fig. 9.15). However, in outer shelf environments below storm wave base, biogenic processes dominate the formation of shell beds (Fig. 9.15).

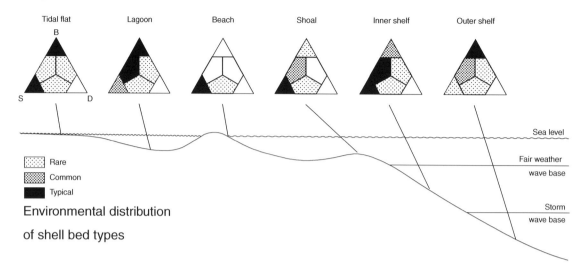

Figure 9.15 Distribution of shell beds along an environmental transect. This shows the expected relative abundances of biogenic, sedimentologic, and diagenetic concentrations, using the ternary scheme outlined in Fig. 9.14, along an onshore–offshore transect in a marine setting dominated by terrigenous sedimentation. Rates of sediment accumulation are assumed to be constant across the transect. In this actualistic model, sedimentologic concentrations decrease in abundance from beach to outer shelf because of diminishing water energy at the seafloor, and biogenic concentrations increase in relative abundance. From Kidwell et al. (1986). Reproduced with permission from the SEPM Society for Sedimentary Geology.

Cold seeps and hydrothermal vents

Another way to develop concentrations of skeletal material in sedimentary environments is through the formation of fossil seep and hydrothermal vent deposits, which are found in both siliciclastic and carbonate settings. As has been discussed in Chapter 3, these deposits have only been recognized for the past 30 years since the recognition of hydrothermal vent and cold seep deposits on the modern seafloor. Cold seep deposits are usually caused by seepage of fluids and gases including oil, methane, and hydrogen sulfide from below to the seafloor (Fig. 9.16). This seepage is through features such as faults, mud diapirism and volcanism, and sand injection related to compaction of sediments and particularly tectonic convergence in subduction zones (Fig. 9.16). Organisms colonizing these seepage sites are not only microbes but also metazoans (Figs. 3.8 and 9.16). These fascinating organisms

include various kinds of bivalves (Fig. 9.17) and vestimentiferan tube worms and also extensive microbial deposits, including precipitation of carbonates. Animals were living at ancient seeps by at least the Silurian and until the Jurassic included various brachiopods, monoplacophoran molluscs, bivalves, and gastropods. After the Jurassic, this earlier fauna was replaced largely by extant representatives of modern bivalves and gastropods. As discussed in Chapter 3, many fossil seeps were originally thought to represent shallow-water carbonate deposits, but we now know they represent cold seeps.

Ancient hydrothermal vent deposits are less common than cold seep accumulations, because they usually do not occur along continental margins and commonly require tectonic accretion of ocean basalts (ophiolites) formed in the deep sea to be preserved. Ancient vents show the occurrence of animals by at least the Silurian. Figure 9.18 shows a reconstruction of a typical Early Jurassic

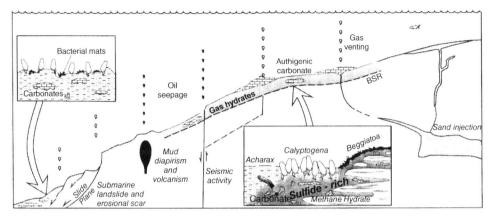

Figure 9.16 Diagram showing how and where hydrocarbon seep accumulations are formed. This schematic portrays the links between fluid migration pathways of seep plumbing, gas hydrate distribution, fluid expulsion features, authigenic carbonate precipitates, and typical seep communities in continental margin areas charged with hydrocarbons. Fluid overpressuring occurs with sediment loading and/or convergence-related deformation to yield high permeability, especially within fault zones and via buoyant mud volcanoes and diapirs. Hydrocarbons are transported via sand injection, through trapping/release of free gas beneath accumulations of methane clathrates, fault conduits, mud diapirism and volcanism, and leakage from sites of submarine slides/tectonic erosion. Methane gas hydrates are ice-like solids formed of methane and water which occur in areas of voluminous methane production under certain temperature and pressure conditions, and the base of such accumulations below the seafloor forms a seismic reflector that parallels the seafloor reflection (bottom simulating reflector; BSR). *Acharax* and *Calyptogena* are bivalves that have chemosymbiotic sulfide-oxidizing bacteria in their gills, and *Beggiatoa* are bacteria that form microbial mats and oxidize hydrogen sulfide. From Campbell (2006). Reproduced with permission from Elsevier.

Figure 9.17 Cross sections of articulated lucinid bivalves from the Lower Miocene Ugly Hill seep deposit. This site occurs in the uplifted East Coast Basin forearc in the Tolaga Group on the eastern North Island of New Zealand. Head of rock hammer for scale. Additional information can be found in Saether et al. (2010). Photograph by Kathleen A. Campbell. Reproduced with permission. (*See insert for color representation.*)

Figure 9.18 Reconstruction of an Early Jurassic hydrothermal vent setting. This schematic is of the Pliensbachian Figueroa hydrothermal vent community (California, USA), which lived on a high-temperature seafloor vent occurring on a mid-ocean ridge or seamount, and illustrates ecological zonation due to temperature variations of vent fluids. A, Black smoker chimney and hydrothermal fluid plume emitting sulfide-rich fluids from which the chimney was precipitated; B, collapsed extinct chimney; C, apron of sulfide talus and metalliferous sediments; D, area of low-temperature diffuse flow with mat of chemosynthetic bacteria; E, pillow lavas formed from basalt erupted onto the seafloor; F, cluster of vestimentiferan tube worms close to high-temperature hydrothermal activity; G, *Anarhynchia* cf. *gabbi* brachiopods attached to sulfide talus and grazing *Francisciconcha maslennikovi* gastropods. From Little et al. (2004), where data sources are indicated. Reproduced with permission from the Cambridge University Press.

hydrothermal vent and its fauna, which formed on a mid-ocean ridge or seamount and is now accreted to southern California.

Summary

These various carbonate deposits represent an important addition to the information that can be accessed from level-bottom environments on the evolution and paleoecology of benthic marine life. Such deposits are commonly outcropping as continuous hard limestones, with organisms such as corals that have large skeletons. Thus, ancient reefs, seeps, shell beds, and hydrothermal vents are not typically studied through collection of bulk samples. Rather, outcrop counts of fossils along linear transects or in a specified surface area are made, and data from point count analyses of thin sections are also collected, to produce quantitative data for paleoecological reconstruction. These various sources of data have provided sets of different yet commonly complementary data to that collected from level-bottom environments. Comparison of the ecological histories of reefs, seeps, and level-bottom environments (e.g., Sheehan 1985; Kiessling et al. 2002; Campbell 2006) is ongoing, and this has provided insights to unraveling the long history of benthic marine ecosystems.

References

Bernecker, M., Weidlich, O. & Flügel, E. 1999. Response of Triassic reef coral communities to sea-level fluctuations, storms and sedimentation: Evidence from a spectacular outcrop (Adnet, Austria). *Facies* 40, 229–280.

Campbell, K.A. 2006. Hydrocarbon seep and hydrothermal vent paleoenvironments and paleontology: Past developments and future research directions. *Palaeogeography, Palaeoclimatology, Palaeoecology* 232, 362–407.

Greene, S.E., Bottjer, D.J., Hagdorn, H. & Zonneveld, J.-P. 2011. The Mesozoic return of Paleozoic faunal constituents: A decoupling of taxonomic and ecological dominance during the recovery from the end-Permian mass extinction. *Palaeogeography, Palaeoclimatology, Palaeoecology* 308, 224–232.

James, N.P. & Wood, R. 2010. Reefs. In James, N.P. & Dalrymple, R.W., *Facies Models 4*. Geological Association of Canada, pp. 421–447.

Kidwell, S.M., Fursich, F.T. & Aigner, T. 1986. Conceptual framework for the analysis and classification of fossil concentrations. *Palaios* 1, 228–238.

Kiessling, W., Flügel, E. & Golonka, J. (eds.). 2002. *Phanerozoic Reef Patterns. Society for Sedimentary Geology (SEPM), Special Publication* **72**.

Little, C.T.S., Danelian, T., Herrington, R.J. & Haymon, R.M. 2004. Early Jurassic hydrothermal vent community from the Fransiscan Complex, California. *Journal of Paleontology* 78, 542–559.

Martindale, R.C., Krystyn, L., Bottjer, D.J., Corsetti, F.A., Senobari-Daryan, B. & Martini, R. 2013. Depth transect of an Upper Triassic (Rhaetian) reef from Gosau, Austria: Microfacies and community ecology. *Palaeogeography, Palaeoclimatology, Palaeoecology* 376, 1–21.

Rowland, S.M. 1984. Were there framework reefs in the Cambrian? *Geology* 12, 181–183.

Saether, K.P., Little, C.T.S. & Campbell, K.A. 2010. A new fossil provannid gastropod from Miocene hydrocarbon seep deposits, East Coast Basin, North Island, New Zealand. *Acta Palaeontologica Polonica* 55, 507–517.

Sheehan, P.M. 1985. Reefs are not so different – They follow the evolutionary pattern of level-bottom communities. *Geology* 13, 46–49.

Stanley, G.D. 2001. Introduction to reef ecosystems and their evolution. *In* Stanley, G.D. (ed.), *The History and Sedimentology of Ancient Reef Systems*, Academic/Plenus Publishers, pp. 1–39.

Wood, R. 2000. *Reef Evolution*. Oxford University Press.

Additional reading

Stanley, G.D. (ed.). 2001 *The History and Sedimentology of Ancient Reef Systems*, Academic/Plenum Publishers.

10 Pelagic ecosystems

Introduction

The study of the ecology of Phanerozoic organisms that lived in the water column above seafloor environments presents different geological challenges for sample collection and environmental interpretation. Pelagic organisms are incorporated into the sedimentary record by descent of their remains to the seafloor where they are deposited with the benthic organisms that had lived there. This presents an added dimension of complexity in reconstructing pelagic ecology. Nevertheless, various interesting paleoecological approaches have been developed, and this is a fascinating field for future research.

Microfossils

The study of organisms that lived in the water column has primarily been done in the past by biostratigraphers. Many biostratigraphic schemes use fossils that swam or drifted in the water column and thus had wide geographic ranges and are very useful for broad interregional correlation. In particular, microfossils provide one of the most important tools for understanding ancient pelagic ecosystems. This is because they occur in great numbers in cores and thus have been used extensively for understanding pelagic paleoecology of ancient marine settings.

One of the common microfossils for biostratigraphy in the Paleozoic and Triassic is the toothlike conodonts. Conodont animals were small chordates that lived in the water column and had a soft body with a biomineralized feeding apparatus made of calcium phosphate, which are the conodont fossils. Typically, only the conodonts are preserved, and there are only a few specimens, preserved by exceptional preservation processes, known of the soft body of the conodont animal. After death, conodonts were deposited in the sediment below and preserved with the remains of benthic organisms that lived there. Attempts have been made to understand where the different conodont animals lived. Figure 10.1 illustrates a Devonian transect from a nearshore and reef environment to deep low oxygen settings which had conodont animals living in the overlying water column. Although the conodonts experienced some transport due to storm processes before final deposition (Fig. 10.1), they occur as assemblages that can be reconstructed into an ecological zonation both vertically in the water column and by distance from the shoreline (Fig. 10.1).

Another group of microfossils that has great utility in biostratigraphy is the coccoliths. These represent some of the most important organisms in modern and ancient oceans and in their role

Paleoecology: Past, Present and Future, First Edition. David J. Bottjer.
© 2016 John Wiley & Sons, Ltd. Published 2016 by John Wiley & Sons, Ltd.

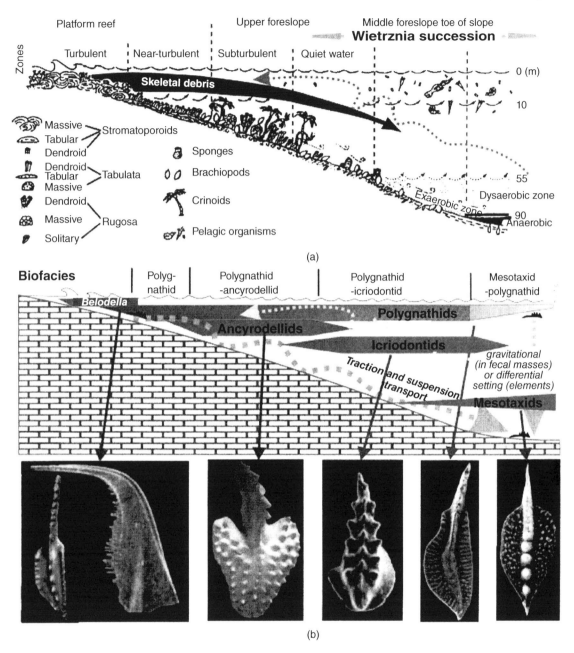

(a)

(b)

Figure 10.1 Conodont paleoecology and biofacies in environments of early Frasnian shallow reef to deep basin storm-influenced settings. Interpretations are from strata in the Holy Cross Mountains at the Wietrznia quarry (Poland). (a) Ecological model and depositional processes for benthic and pelagic organisms of the Wietrznia succession and adjacent environments. (b) "Standard" biofacies model for Frasnian conodonts, determined from a synthesis of conodont distribution and sedimentary data from a variety of examples, with schematic presentation of postmortem conodont deposition, burial, and transport in the storm-affected regime; an upslope redistribution by storm surge action is also shown; representative conodonts for the biofacies are illustrated. From Vierek and Racki (2011). Reproduced with permission from Elsevier.

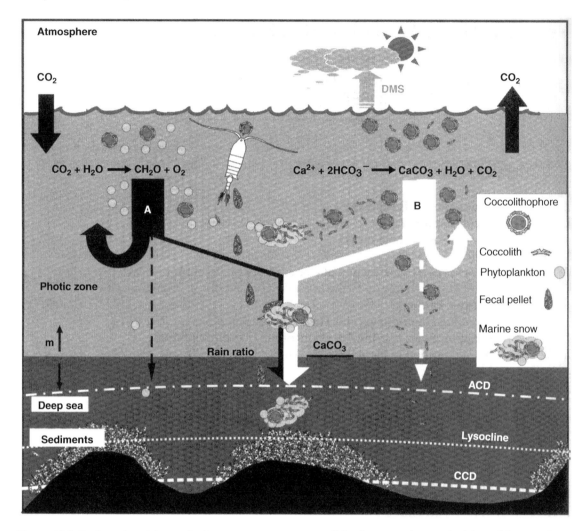

Figure 10.2 Ecology, deposition, and role in biogeochemical cycles of coccolithophores. Coccolithophores are protists that through their production of calcium carbonate coccoliths play a key role in global carbon cycling (Fig. 1.4). Although they thrive in the photic layer of the world's ocean, coccolithophores actively participate in gas exchange (CO_2; dimethyl sulfide – DMS) between seawater and the atmosphere and the export of organic matter and carbonate to deep-oceanic layers and deep-sea sediments. They are the main actors of the carbonate counterpump (B), which, through the calcification reaction, is a short-term source of atmospheric CO_2. Via the ballasting effect of their coccoliths on marine snow, coccolithophores are also a main driver of the organic carbon pump (A), which removes CO_2 from the atmosphere. Thus, organic and carbonate pumps are tightly coupled through coccolithophore biomineralization. Ultimately, certain types of coccoliths particularly resistant to dissolution are deposited at the seafloor above the CCD, where they have built a remarkable fossil archive for the last 220 million years. The three main carbonate dissolution horizons are depicted: ACD, aragonite compensation depth; lysocline (complete dissolution of planktic foraminifera); and CCD, calcite compensation depth. From De Vargas et al. (2007), where data sources are indicated. Reproduced with permission from Elsevier.

within ocean ecosystems (Fig. 10.2). Coccoliths are calcite platelets several microns in greatest dimension that form a composite exoskeleton for the coccolithophore. Coccolithophores are protists that are primary producers which first evolved in the Early Mesozoic and precipitate an estimated half of all calcium carbonate in modern oceans and are therefore very important in the carbon cycle (Fig. 1.4). Coccolithophores live in the photic zone and are consumed by zooplankton, which package them into fecal pellets. Most coccoliths become part of the stratigraphic record by drifting down to the seafloor as fecal pellets and larger aggregates that are termed marine snow (Fig. 10.2). Coccoliths are very useful in understanding the ecology of ancient oceans as

well as how the carbon cycle was working. They are a major component of the calcareous ooze that covers the ~35% of the modern ocean seafloor which is above the depth of complete calcite dissolution (calcite compensation depth, CCD) (Fig. 10.2) and the resulting deep ocean sedimentary record.

Integrated studies

Complete integrated studies of ancient pelagic ecology are difficult to do and are not very common. Holistic studies that attempt to integrate pelagic and benthic ecosystems, such as for the Devonian of the

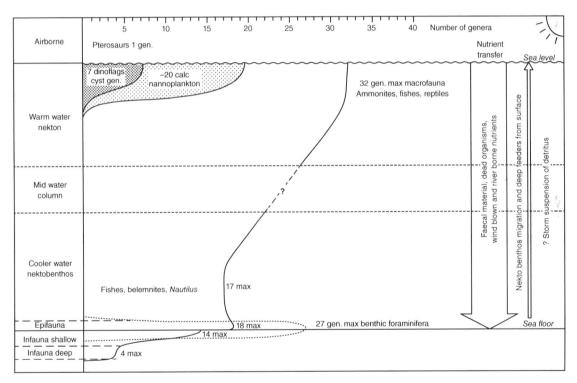

Figure 10.3 Distribution of organisms within pelagic and benthic ecosystems of a Middle Jurassic epicontinental sea. This is the Peterborough Member Sea, from a study of the biota of the Peterborough Member of the Oxford Clay Formation (United Kingdom). Water depth is not known with great accuracy. The water and sediment column is divided into six realms: an upper,

warm water nektonic realm; a possible mid-water column; cooler water nektobenthonic realm; an epifaunal realm; shallow infaunal realm; and deep infaunal realm. The number of genera inhabiting each realm is indicated. From Martill et al. (1994). Reproduced with permission from the Geological Society.

Holy Cross Mountains (Fig. 10.1), are less common, largely because a variety of taphonomic processes need to have been operating to preserve the broad spectrum of organisms originally inhabiting an ancient environment. Another example of this kind of study comes from the Jurassic Oxford Clay of Great Britain, where extremely fossiliferous outcrops with abundant preservation of benthic and pelagic

fossils allow such determinations. Figure 10.3 shows this fossil availability as the reconstructed distribution of genera through various intervals from the seafloor through the overlying water column and the kinds of macrofossils and microfossils that existed. This extensive fossil record is of great utility and has been used to determine trophic webs for different aspects of this seaway, such as that for the upper

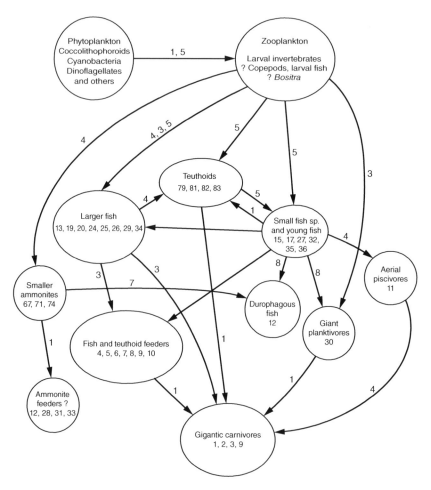

Figure 10.4 Jurassic epicontinental sea pelagic food web. This is illustrated as the trophic relationships for genera, or groups of genera, interpreted as having fed largely within the upper parts of the water column, from a study of the biota of the Jurassic Peterborough Member of the Oxford Clay Formation (United Kingdom). Numbers within the circles represent individual taxa, and numbers on links between the circles represent subjective confidence levels interpreted for each of the links, from 1 to 8, where 1 is high and 8 is low, defined in Martill et al. (1994). Note that the gigantic carnivores are marine reptiles – pliosaurs and crocodiles. From Martill et al. (1994). Reproduced with permission from the Geological Society.

water column and hence the pelagic ecology of this Jurassic epicontinental sea (Fig. 10.4). In contrast to Cenozoic and modern marine environments, marine reptiles, as giant carnivores, rather than marine mammals, were prevalent in these Mesozoic settings.

Macrofossils

Marine reptiles were also common in the Cretaceous, when sea level was high, creating extensive epicontinental seaways such as the Western Interior Seaway of North America, which stretched at times from the Gulf of Mexico to the Arctic Ocean. Sediments in Cretaceous seas contain abundant benthic faunas but also are notable for their common occurrence of the organisms that lived in the overlying water column. These organisms include Mesozoic marine reptiles as well as abundant ammonites. Figure 10.5 illustrates typical marine reptiles such as plesiosaurs, ichthyosaurs, and mosasaurs that lived in pelagic environments of Cretaceous seas, accompanied by abundant ammonites.

Externally shelled cephalopods lived from the seafloor to the ocean surface and were abundant in marine environments from the Ordovician through the Mesozoic. These molluscs, including nautiloids and ammonoids (such as ammonites), are thus one of the most common marine fossils and are widely used for biostratigraphic studies. Because all but the modern *Nautilus* are extinct, paleoecology of this once abundant group has posed additional

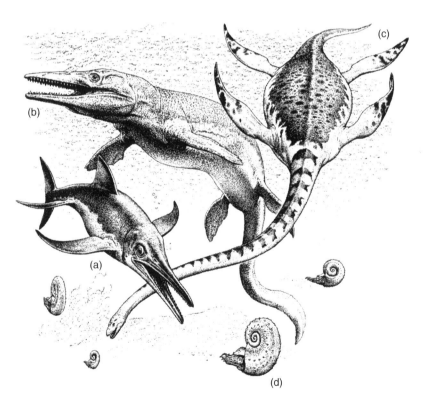

Figure 10.5 Typical inhabitants of pelagic habitats in Cretaceous seas. These include marine reptiles such as (a) ichthyosaurs (*Platypterygius*), (b) mosasaurs (*Tylosaurus*), and (c) plesiosaurs (*Elasmosaurus*), as well as (d) ammonite cephalopod molluscs. From Fastovsky and Weishampel (2009). Illustrated by John Sibbick. Reproduced with permission from the Cambridge University Press.

challenges, leading to a variety of approaches to understanding how they lived.

For example, in the Western Interior Seaway of the Cretaceous, abundant ammonites are found with a variety of intriguing morphologies. Figure 10.6 shows a study that synthesized biostratigraphic, paleobiogeographic, and sedimentologic data for Western Interior ammonites to make conclusions on their lifestyles and environmental distribution in this Cretaceous seaway. This study provides a fairly detailed look at the ecology of these fascinating organisms during this time. Important faunal components are ammonites with a broad distribution throughout the seaway, and these morphotype groups (8, 10, 14, 16, 17; Fig. 10.6), may have largely been drifting in pelagic ecosystems. The remaining morphotype groups depicted in Fig. 10.6 were typically more mobile and commonly tied to seafloor environments.

Other approaches have attempted to study ammonoid ecology from an analysis of their shell morphology. For example, Fig. 10.7 shows that the major planispiral ammonoid morphologies can be distributed from actively swimming to drifting in the water column to swimming along and above the seafloor. This kind of analysis begins with a synthesis of morphological, paleoecological, and experimental studies on ammonoids in which ammonoid morphology is portrayed on a triangular diagram known as Westermann Morphospace (Fig. 10.7). Work done with Kathleen Ritterbush shows that morphological measurements can then allow placement of each measured specimen within the triangular diagram, which portrays the prevalence of each ammonoid life habit. The analysis presented in Fig. 10.8 shows that in the Middle Triassic, ammonites occupied most of these various life habits, but in the Early Jurassic, after the end-Triassic mass extinction,

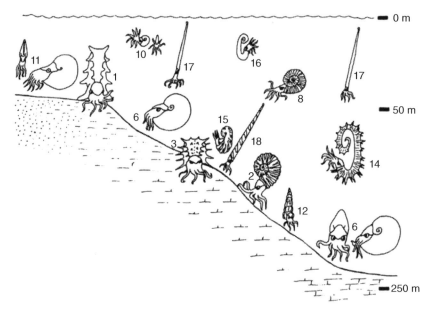

Figure 10.6 Habitats of ammonites of selected morphotype groups found in the Cretaceous North American Western Interior Greenhorn Sea. This synthesis is from samples ranging from Montana to New Mexico and Texas. Each numbered sketch represents the overall morphology for a specific morphotype group; definition and discussion of these morphotypes are found in Batt (1989). Seafloor to sea surface environments for each morphotype group were determined based on biostratigraphic and paleobiogeographic distributions as well as sedimentologic data. From Batt (1989). Reproduced with permission from the SEPM Society for Sedimentary Geology.

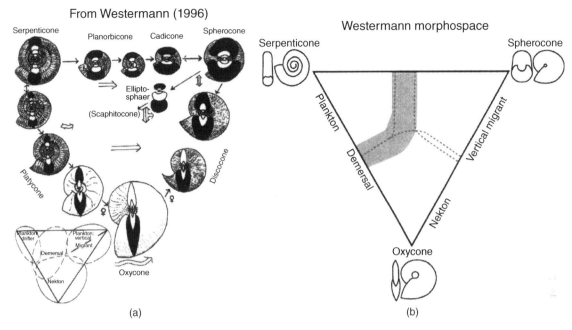

Figure 10.7 Paleoecology of planispiral ammonoids. This is portrayed as Westermann Morphospace, where planispiral ammonoids of different shapes are associated with different hypothetical life modes, as synthesized by Westermann (1996). (a) Summary of ammonoid mobility for common planispiral shell shapes that grade between three forms – serpenticone, sphaerocone, and oxycone. Hypothetical life modes are indicated by the inset triangle. Each illustration includes a side view of the outer shell and a cross section through the whorls, superimposed. (b) Westermann Morphospace, based on (a). The three components of shell shape (exposure of the umbilicus, overall inflation, and whorl expansion) for ammonite specimens dictate their placement within this ternary diagram. Serpenticones (high umbilical exposure, low overall inflation, low whorl expansion) were incapable of directed swimming and plot in the plankton field. Sphaerocones (low umbilical exposure, high inflation, and low whorl expansion) are interpreted to have moved up and down in the water column and thus plot in the vertical migrant field. Oxycones (low umbilical exposure, low inflation, and high whorl expansion) would have been efficient swimmers and thus plot in the nekton field. Platycones and planorbicones were mobile, moving along the seafloor, and thus plot in the demersal field. Details on terminology of ammonoid morphology can be found in Ritterbush and Bottjer (2012). From Ritterbush and Bottjer (2012). Reproduced with permission from the Cambridge University Press. (*See insert for color representation.*)

the ammonite faunas are characterized primarily by drifters.

Belemnites are another extinct group of cephalopods that were common in Jurassic and Cretaceous pelagic marine environments. They lived like squid (Fig. 10.9) and had internal skeletons largely made of calcite, termed the rostrum. These calcite skeletons are typically how they occur as fossils, and abundant accumulations are an indication that they were typically common in pelagic ecosystems (Fig. 10.9). This phenomenon of belemnite-rich shell beds is one that has intrigued paleoecologists. Because belemnite rostra have the shape of bullets, these belemnite shell beds, where many rostra can be weathered out on the surface, are commonly called belemnite battlefields. Belemnite battlefields accumulated in a variety of ways, which provide evidence on belemnite behavior and

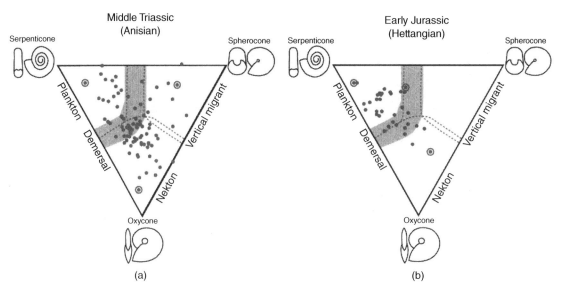

Figure 10.8 Comparison of ammonoid shell shapes and hypothetical life modes in Westermann Morphospace. (a) Plot of Middle Triassic ammonoids of Nevada include each major morphotype. Each point represents the largest measurable specimen of each species ($N = 85$) in the collections. (b) Earliest Jurassic ammonoids of Nevada include a comparatively limited variety of shell shapes. Each point represents the largest measurable specimens of each species ($n = 35$). All specimens were measured from monographs with calipers. Measurement error is about 1.7 mm. When repeat measurements of a specimen are plotted together in Westermann Morphospace, the plotted points partially overlap; the error is not great enough to cause noticeable change in the position of points. Specimens corresponding to circled data points are illustrated in Ritterbush and Bottjer (2012). From Ritterbush and Bottjer (2012). Reproduced with permission from the Cambridge University Press. (*See insert for color representation.*)

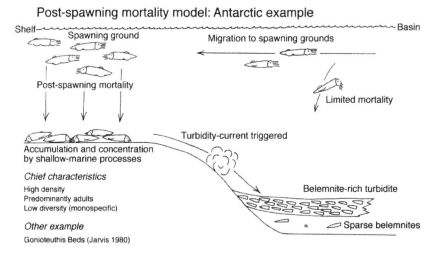

Figure 10.9 Conceptual model for deposition of belemnite battlefields. These are derived from observations on the Jurassic–Cretaceous Fossil Bluff Group of Alexander Island, Antarctica. Battlefield shell beds characterized by a primarily monospecific high-density accumulation of adult belemnites are interpreted to occur due to postspawning mortality in the overlying water column. Parts of such accumulations can then be transported by turbidity currents to deeper marine areas, forming battlefield shell beds in those settings. From Doyle and MacDonald (1993). Reproduced with permission from Taylor and Francis.

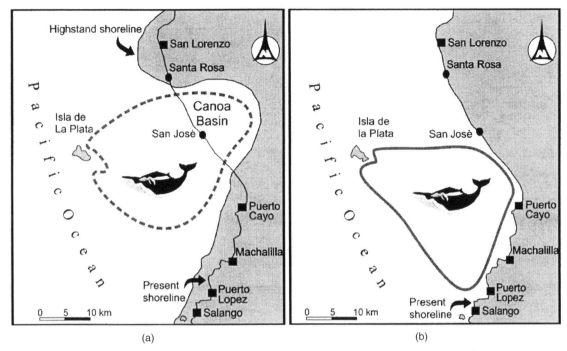

Figure 10.10 Humpback whale breeding areas off the Ecuador coast: (a) during the warm phases of Late Pliocene–Pleistocene (delimited by a dashed line) and (b) at the present (delimited by a continuous line). From Bianucci et al. (2006). Reproduced with permission from Elsevier.

pelagic ecology. One mechanism of accumulation is hypothesized to have been postspawning mortality, as occurs for many modern squid, which would lead to preferential accumulation of belemnites on the seafloor (Fig. 10.9). These belemnite accumulations might then be subject to transport as turbidites (Fig. 10.9). They might also be concentrated by predators that ingest large numbers of belemnites and then regurgitate the belemnite shell to produce an accumulation on the seafloor.

Breeding behavior of other pelagic organisms can also be determined through studies of the fossil record. Whale barnacles live on and in the skin of whales, and their fossils consist of several large calcium carbonate plates. Although typically rare as fossils, they are found relatively commonly within shell beds otherwise dominated by molluscs in Plio-Pleistocene marine deposits of the Canoa Basin in Ecuador (Fig. 10.10). These fossil whale barnacles, together with cetacean remains, are interpreted to

indicate that modern humpback whale breeding grounds off of Ecuador have existed there since at least the Plio-Pleistocene (Fig. 10.10).

Summary

Holistic studies of fossil pelagic ecosystems require expertise in micropaleontology, vertebrate paleobiology, and invertebrate paleontology. Sample availability of larger vertebrate and invertebrate fossils is typically limited to epicontinental seas and ancient ocean margins, with a rich microfossil record potentially available for most pelagic settings. With the increase in studies of ancient pelagic ecosystems, an integrated understanding of the evolutionary paleoecology of benthic and pelagic systems and their interactions through time, particularly in ocean margin and epicontinental sea settings, will become possible.

References

Batt, R.J. 1989. Ammonite shell morphotype distributions in the Western Interior Greenhorn Sea and some paleoecological implications. *Palaios* 4, 32–42.

Bianucci, G., Di Celma, C., Landini, W. & Buckeridge, J. 2006. Palaeoecology and taphonomy of an extraordinary whale barnacle accumulation from the Plio-Pleistocene of Ecuador. *Palaeogeography, Palaeoclimatology, Palaeoecology* 242, 326–342.

De Vargas, C., Aubry, M.-P., Probert, I. & Young, J. 2007. Origin and evolution of coccolithophores: From coastal hunters to oceanic farmers. *In* Falkowski, P.G. & Knoll, A.H., *Evolution of Primary Producers in the Sea*, Elsevier, pp. 251–285.

Doyle, P. & MacDonald, D.I.M. 1993. Belemnite battlefields. *Lethaia* 26, 65–80.

Fastovsky, D.E. & Weishampel, D.B. 2009. *Dinosaurs: A Concise Natural History*. Cambridge University Press.

Martill, D.M., Taylor, M.A., Duff, K.L., Riding, J.B. & Bown, P.R. 1994. The trophic structure of the biota of the Peterborough Member, Oxford Clay Formation (Jurassic), UK. *Journal of the Geological Society, London* 151, 173–194.

Ritterbush, K.A. & Bottjer, D.J. 2012. Westermann Morphospace displays ammonoid shell shape and hypothetical paleoecology. *Paleobiology* 38, 424–446.

Vierek, A. & Racki, G. 2011. Depositional versus ecological control on the conodont distribution in the Lower Frasnian fore-reef facies, Holy Cross Mountains, Poland. *Palaeogeography, Palaeoclimatology, Palaeoecology* 312, 1–23.

Additional reading

Ellis, R. 2003. *Sea Dragons: Predators of the Prehistoric Oceans*. University Press of Kansas.

Everhart, M.J. 2005. *Oceans of Kansas: A Natural History of the Western Interior Sea*. Indiana University Press.

Falkowski, P.G. & Knoll, A.H. (eds.). 2007. *Evolution of Primary Producers in the Sea*. Academic Press, Elsevier.

Landman, N.H., Davis, R.A. & Mapes, R.H. (eds.). 2007. *Cephalopods Present and Past: New Insights and Fresh Perspectives*. Springer, Dordrecht.

Westermann, G.E.G. 1996. Ammonoid life and habitat. *In* Landman, N., Tanabe, K. & Davis, R.A. (eds.), *Ammonoid Paleobiology. Topics in Geobiology* 13, 607–707. Plenum Publishing Corporation.

11 Terrestrial ecosystems

Introduction

Fossils from Phanerozoic terrestrial ecosystems are diverse, including plants, their pollen, vertebrates, and invertebrates. Thus, in particular, a knowledge of paleobotany, palynology, and vertebrate paleontology is crucial for undertaking paleoecological analyses in the terrestrial realm. An understanding of microbial structures, as well as bioturbation structures made by vertebrates and invertebrates in sediment, and evidence of herbivory on leaves and predation on bones has also become important in revealing additional paleoecological information. Fossils from these sedimentary systems are preserved under more normal taphonomic conditions as well as in depositional environments producing exceptional preservation.

Development of ecosystems on land

Precambrian terrestrial ecosystems consisted of cyanobacteria and possibly also algae and later fungi, which likely contributed to the formation of cryptobiotic crusts and soils. The earliest land plants are from the Ordovician, and vascular plants, which are able to conduct water throughout the plant, had evolved by the Silurian. Arthropods had colonized land by at least the Ordovician. One of the best known early ecosystems that included vascular plants and arthropods is from the Lower Devonian Rhynie chert, a lagerstätten which is found in Scotland (Fig. 11.1). This ecosystem included seven species of small vascular plants, reaching no greater than several tens of centimeters above the surface (Fig. 11.1), which was also inhabited by bacteria, algae, fungi, and lichens. Animals included arachnids, mites, centipedes, springtails, and crustacean arthropods as well as an extinct arthropod (Fig. 11.1), adding to ecological complexity.

After the colonization of land by arthropods came the vertebrates. Devonian freshwater ecosystems were inhabited by fish. By the Late Devonian, tetrapods such as *Acanthostega* and *Ichthyostega* were living on the edges of freshwater environments and interacting as part of terrestrial ecosystems. This evolution from fish in freshwater environments to tetrapods is exemplified by the Late Devonian sarcopterygian fish *Tiktaalik*, which represents an intermediate form on its way to primarily inhabiting land (Fig. 11.2). By the early Carboniferous, amphibians were playing a significant role in marginal aquatic environments (Fig. 3.12).

As discussed in Chapter 2, the evolution of lignin in vascular plants by the middle of the Paleozoic provided structural support for plants to grow to significant heights. Thus, the Carboniferous was

Paleoecology: Past, Present and Future, First Edition. David J. Bottjer.
© 2016 John Wiley & Sons, Ltd. Published 2016 by John Wiley & Sons, Ltd.

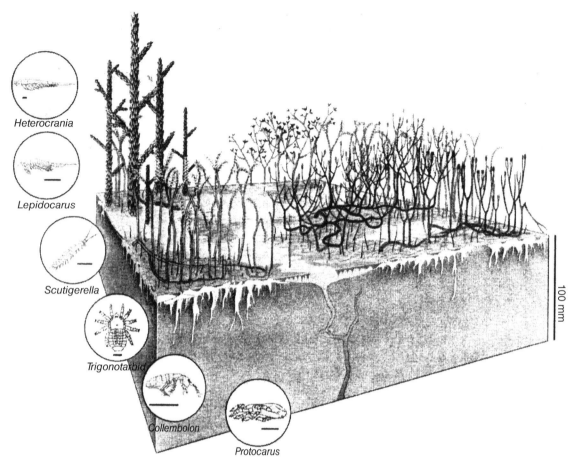

Heterocrania

Lepidocarus

Scutigerella

Trigonotarbid

Collembolon

Protocarus

100 mm

Figure 11.1 Early Devonian terrestrial ecosystem with vascular plants and arthropods. This is reconstructed from an analysis of the Lower Devonian Rhynie chert. The most common vascular plants were *Rhynia* and *Asteroxylon* (foreground). A variety of small arthropods, shown on the left and bottom center, lived in the water and in and on the plants (scale bars, 100 μm). *Heterocrania* is a member of the extinct euthycarcinoid arthropods; *Lepidocarus* is a crustacean; *Scutigerella* is a centipede; a trigonotarbid is an arachnid; *Collembolon* is a springtail, which is a Hexapoda, which also includes insects; and *Protocarus* is a mite. This lagerstätten was preserved by silicification from adjacent silica-rich hot spring waters in a setting similar to modern Yellowstone Park (United States) geysers and hot springs. Illustration by Simon Powell. From Benton and Harper (2009). Reproduced with permission from John Wiley & Sons.

a time of extensive forest development (Fig. 2.6) and deposition of coals. These early forests with very tall trees represent the earliest rain forests from tropical environments. Coals and associated strata have a broad variety of plant fossils representing the organisms which lived in these coal forests. In one particularly remarkable example, a Pennsylvanian coal swamp forest was drowned by a rapid fault movement that caused subsidence and flooding of the forest (Fig. 11.3). A reconstruction of environmental conditions during this time is shown in Fig. 11.4. The tectonic event preserved

Early amphibian

Tiktaalik

Lobe-finned fish

Figure 11.2 Vertebrates evolve to inhabit terrestrial ecosystems. This shows the habitat and evolution of the Late Devonian sarcopterygian fish *Tiktaalik*. On left the skeleton is shown below, with a reconstruction of *Tiktaalik* shown above in a shallow freshwater environment marginal to land where it lived. *Tiktaalik* was intermediate between lobe-finned fishes and amphibians and had a flat, crocodile-like skull with eyes and nostrils on the upper surface. Its fins were intermediate in form between those of lobe-finned fishes and amphibians, with shoulder joints and finger-like bones which would have permitted it to prop up the front of its body. In shallow water, this would have allowed it to survey its surroundings. Three bones in its front fin are portrayed on the right in light and dark gray which show homologies with bones of lobe-finned fishes and early amphibians. *Tiktaalik* ranged from 2 to 3 m in length. Illustration by Zina Deretsky, photo by Ted Daeschler. From Stanley (2008). (*See insert for color representation.*)

many elements of the forest *in situ* and has allowed paleoecological study of the spatial distribution of forest components. This forest was dominated by lycopsids that were 40 m or more tall, with a subcanopy dominated by tree ferns, also including cordaitalean, pteridosperm, and sphenopsid small trees and shrubs (Fig. 11.4).

Post-Paleozoic terrestrial ecosystems

In terrestrial environments, vertebrate bones can be deposited in bedded accumulations analogous to shell beds in marine environments. A bonebed can be defined as a skeletal accumulation that contains the remains of two or more individuals, although typically there are many more. Detailed studies of bonebeds have been made and a variety of paleoecological evidence can be obtained from them, including ecological associations, life histories, and behavior. Like shell beds, bonebeds are formed by both biotic and physical mechanisms. Physical mechanisms include processes involving water or wind where transport to a particular location, such as a strandline, concentrates bones. They can also involve reductions in sedimentation rate and reworking of existing sediments to form bone concentrations. A bonebed formed by

Figure 11.3 Upright stump of the lycopsid *Sigillaria* rooted into the top of a coal seam (Herrin Coal) which is Middle Pennsylvanian (Desmoinesian) in age. Tree stump, located at the end of the extended tape measure, is encapsulated in tidal rhythmites (interlaminated silt and silty mud) and is preserved as described in Fig. 11.4. White-colored "dust" on coal mine walls and ceiling is limestone dust used for coal dust suppression. Man at left (William A. DiMichele, Smithsonian Institution) for scale. Photograph by Scott D. Elrick. Reproduced with permission. (*See insert for color representation.*)

Figure 11.4 Preservation of a Carboniferous tropical rain forest. This illustrates a model for the origin of a Middle Pennsylvanian fossil forest found on the top of the Herrin (No. 6) Coal of Illinois (United States) (Fig. 11.3). (a) Onset of differential subsidence and its effects on mire (swamp) community structure. The Herrin peat mire was formed from a tall rain forest with a spatially heterogeneous variety of plants, dominated by lycopsids and tree ferns. Subsidence on the eastern side of the fault resulted in submergence of the mire, and there is a reduction in diversity and loss of shrubs and ground cover in this easterly direction. The mire became peat as shown in the vertical panels and, with continued deposition and compaction, later became the Herrin Coal (Fig. 11.3). (b) Abrupt subsidence of the eastern side through activity of the subsurface fault resulted in submergence and formation of estuarine environments including tidal channels. This led to preservation in estuarine muds of components of the drowned forest (Fig. 11.3), such as stumps and trunks, which underwent none (autochthonous) to some (parautochthonous) transport. After this event, an ephemeral forest developed along the margin of the estuary, typically termed a gallery forest. The spatial relationships of the drowned mire forest were studied from the roof of underground mines in the Herrin Coal at two localities, Riola (RA) and Vermillion Grove (VG). From DiMichele et al. (2007). Reproduced with permission from the Geological Society of America.

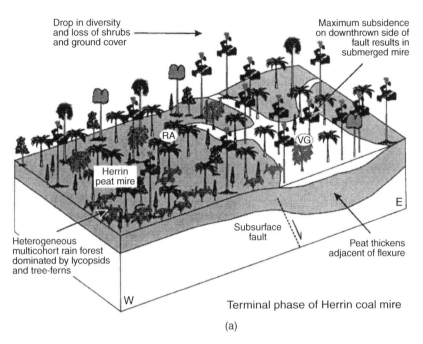

Drop in diversity
and loss of shrubs
and ground cover

Maximum subsidence
on downthrown side of
fault results in
submerged mire

RA

VG

Herrin
peat mire

E

Heterogeneous
multicohort rain forest
dominated by lycopsids
and tree-ferns

Subsurface
fault

Peat thickens
adjacent of flexure

W

Terminal phase of Herrin coal mire

(a)

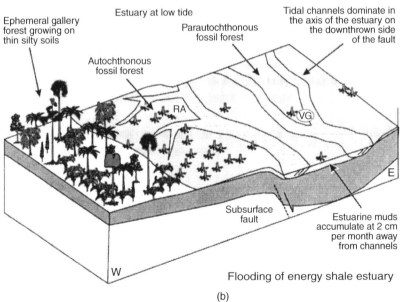

Ephemeral gallery
forest growing on
thin silty soils

Estuary at low tide

Parautochthonous
fossil forest

Tidal channels dominate in
the axis of the estuary on
the downthrown side
of the fault

Autochthonous
fossil forest

RA

VG

E

Subsurface
fault

Estuarine muds
accumulate at 2 cm
per month away
from channels

W

Flooding of energy shale estuary

(b)

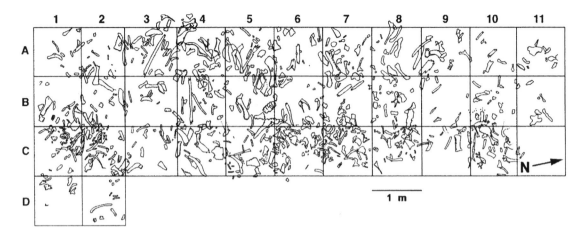

Figure 11.5 Quarry map of an Upper Cretaceous bonebed which contains the remains of a large number of dinosaurs that likely died in a mass kill. This is Bonebed 43 from the Dinosaur Park Formation (upper Campanian) in Dinosaur Provincial Park, Alberta (Canada), which is dominated by the ceratopsian *Centrosaurus*. Deposition occurred in a paleochannel, where the *Centrosaurus* bones were deposited after these ornithiscians drowned during flooding on the alluvial plain. Also included are shed teeth of the tyranosaurid *Albertosaurus libratus*, which likely scavenged the mass kill at the original site of death. Size range of bones indicates that they are from ceratopsians ranging from juvenile to adult. From Ryan et al. (2001), where data sources are indicated. Reproduced with permission from the SEPM Society for Sedimentary Geology.

biotic mechanisms involves biological processes as a significant component in the formation of the concentration. For example, as discussed in Chapter 4, bone concentrations such as those found at the La Brea Tar Pits formed due to behavioral activity related to trapping of organisms in ponds of freshwater that also contained concentrations of tar. Another example of behavior involved in the formation of bonebeds can be seen in Fig. 11.5, which shows a map of a portion of Bonebed 43 from Upper Cretaceous strata in Dinosaur Provincial Park, Alberta. This is one of several bonebeds in the same stratigraphic interval that are interpreted to have been deposited in a river channel and floodplain environment. The bonebed assemblage is dominated by the ceratopsian *Centrosaurus* and individuals of varying ages are present. Bonebed 43 and associated bonebeds are thus interpreted to have formed from mass kills of large herds of these ceratopsians during floods. If preservation

is excellent, a bonebed can also be thought of as a concentration lagerstätten.

Bone concentrations can also result from nesting behavior. Dinosaurs are well known for their nesting sites where accumulations of eggs and very young dinosaurs can be found, as depicted in Fig. 11.6, which shows eggs and a neonate of the saurischian theropod *Gigantoraptor erlianensis*. Similarly, an Upper Cretaceous accumulation of juvenile dinosaur bones from a *Protoceratops* nest in Mongolia is shown in Fig. 11.7. Such accumulations (Fig. 11.7) have allowed a good understanding of the early development of these ceratopsians, which show growth somewhat beyond the hatching stage, implying parental care was involved after birth.

Some dinosaur nesting sites can be spectacular sources of a variety of information. Auca Mahuevo, in Patagonia, was discovered by Luis Chiappe, Lowell Dingus, and Rodolfo Coria and was an enormous Late Cretaceous titanosaurid sauropod nesting ground more than a square kilometer in

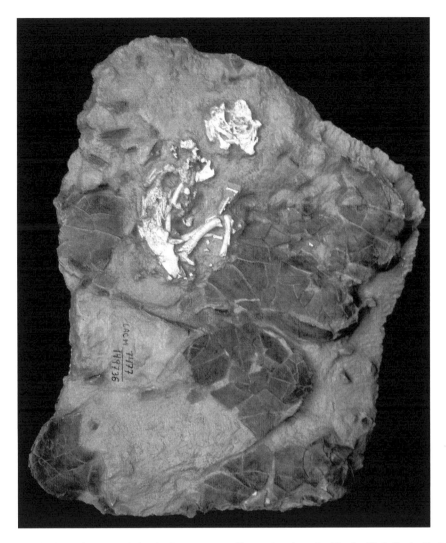

Figure 11.6 Eggs (black) and neonate (white) of *Gigantoraptor erlianensis*, an oviraptorid theropod dinosaur. This specimen (LACMA 7477/149736) is from Sanlimiao in Henan Province, China, and was found in the Upper Cretaceous (Maastrichtian) Zoumagang Formation deposited in the Xixia Basin. Embryo is 25 cm from top to bottom. Photograph by Luis Chiappe. Reproduced with permission. (*See insert for color representation.*)

area. This site contains abundant eggs, including eggs with the skeletons of embryos inside, and some of the embryo fossils are associated with impressions of their skin (Fig. 11.8). Egg clutches are apparent over broad surfaces at Auca Mahuevo, demonstrating that this was a large colonial nesting ground (Fig. 11.9). No adult fossils are present in the deposit, implying that, after constructing the nests and laying the eggs, females probably left the nesting ground. These nesting grounds were located in a flood basin where occasionally rivers overtopped their banks and flooded the nesting grounds, depositing a layer of sediment that buried the eggs and killed the developing embryos. The

Figure 11.7 Dinosaur nest bone concentrations. This depicts the bones of juvenile *Protoceratops* concentrated in a nest from the Upper Cretaceous of Mongolia. Inset: reconstruction of one of the hatchlings as it would have appeared in life. Individual juveniles are no longer than 23 cm. From Fastovsky and Weishampel (2009). Reproduced with permission from Cambridge University Press.

resulting deposit has exceptional fossil preservation and can be considered a conservation lagerstätten.

Dinosaur tracks and trackways (Figs. 5.9 and 5.10) are also a significant source of information on social behavior (Fig. 5.11). At a site along the Paluxy River in Texas, 12 trackways of sauropods have been found, and these are interpreted to indicate that this was a herd that was walking across an Early Cretaceous mud flat environment. Associated with several of these sauropod trackways are theropod trackways, which have been shown to have formed at approximately the same time as the sauropod trackways (Fig. 11.10). From this association, it has been hypothesized that these theropods may have been stalking the sauropods, but there is no additional evidence to confirm this inference.

One of the most important phenomena in evolution to occur was the evolution of flight. Insects had evolved the ability to fly by the Carboniferous (Fig. 2.6). The first flying vertebrates were pterosaurs (Fig. 3.4), and the earliest examples are from the Late Triassic. As indicated by the early bird *Archaeopteryx*, by the Late Jurassic, birds were capable of flight. Various hypotheses requiring different environmental and ecological scenarios have been proposed for the evolution of flight from theropod dinosaurs to birds (Fig. 11.11). With the development in each group of the ability to fly, terrestrial ecosystems would have experienced significant changes in ecospace utilization that would modify and expand trophic webs and produce new evolutionary pathways.

Mobile animals including flying insects and birds have been instrumental in the evolutionary

(a)

(b)

(c)

(d)

(e)

Figure 11.8 Late Cretaceous titanosaur nesting ground at the Auca Mahuevo locality of Patagonia, Argentina. (a) Titanosaur skull (fossil), (b) reconstructed skull, (c) titanosaur skin (fossil) impressions, (d) reconstructed egg/embryo, and (e) schematic field of nests. Scale: the greatest diameter of eggs is 12–14 cm. From Fastovsky and Weishampel (2009). Reproduced with permission from the Cambridge University Press.

and ecological success of the flowering plants. Angiosperms first appear in the fossil record in the Late Jurassic and today are the most diverse land plant group (Fig. 3.12). Angiosperms have evolved other anatomical properties along with flowers that have allowed them to dominate many of Earth's terrestrial ecosystems. One of these is the structure of the leaves. Photosynthesis requires light, water, and CO_2, and the rate at which photosynthesis can proceed is influenced by the rates at which water

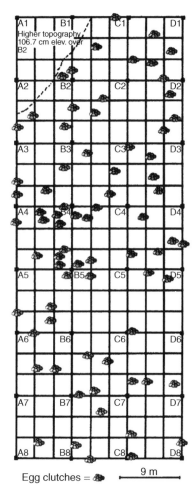

Egg clutches = 🥚   ⊢——— 9 m ———⊣

Figure 11.9 Map of egg clutch distribution on a flat area at Auca Mahuevo. This map shows a high concentration of eggs present at this site. Maps like this one provide evidence on the colonial nesting behavior of sauropod dinosaurs. From Chiappe and Dingus (2001).

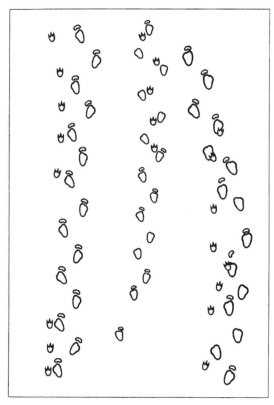

Figure 11.10 Trackways demonstrating that dinosaurs exhibited herding behavior. These footprints, preserved along over 30 m of the Paluxy River bed in Texas, show three sauropod trackways, each overlapped by theropod trackways (bird-like footprints). Such trackways are evidence that sauropods traveled in herds. From this association of trackways, it has also been suggested that these large theropods may have been stalking the sauropods. From Chiappe and Dingus (2001).

can be delivered to leaves. This rate can be inferred by examination of the density of veins in leaves, with increased density of veins providing the ability for higher rates of water delivery, and hence higher photosynthetic rates. Examination of vein density from fossil leaves shows that early angiosperms had vein function similar to nonangiosperm plants (Fig. 11.12). However, angiosperms significantly increased their vein density in the Early Cretaceous and then again at the end of the Cretaceous and the beginning of the Cenozoic, where modern levels are found (Fig. 11.12). The heightened vein density and thus photosynthetically more active leaves are interpreted to have given angiosperms a competitive advantage over gymnosperms and ferns in many terrestrial environments. This in turn would have led to ecological restructuring that affected biogeochemical cycles as well as rainfall and fire regimes.

During warm intervals in Earth's history, there was abundant forest growth at the poles.

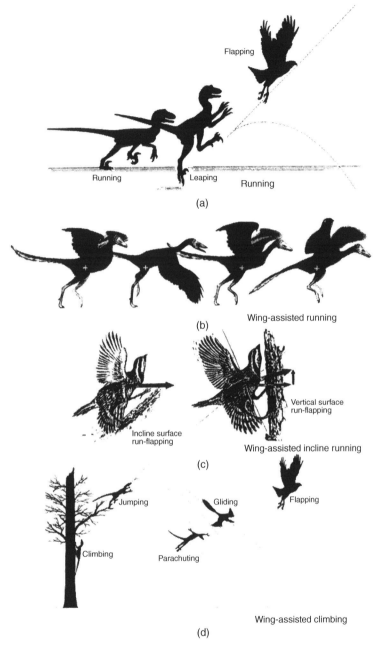

Flapping

Running   Leaping   Running

(a)

Wing-assisted running

(b)

Incline surface
run-flapping

Vertical surface
run-flapping

Wing-assisted incline running

(c)

Jumping   Gliding   Flapping

Climbing   Parachuting

Wing-assisted climbing

(d)

Figure 11.11 Several models for the origin of avian flight involving various functional, environmental, and ecological parameters. Additional hypotheses on the evolution of avian flight can be found in Heers and Dial (2012). (a) The cursorial model involves a running theropod leaping into the air to become an active flier without an intervening gliding state. (b) Wing-assisted running is a modified version of the cursorial model, shown here with a postulated takeoff sequence for *Archaeopteryx*, where running speed increases with thrust generated by the wings so that it begins to take off. (c) Wing-assisted incline running is illustrated by observations of modern partridges as they climb a steep slope by flapping their wings to aid traction, generating a force (arrow) perpendicular to the plane of wing movement. (d) Wing-assisted climbing is illustrated with a progenitor of *Archaeopteryx* climbing assisted by wings, jumping, then parachuting from trees using gravity as the source of power, then beginning to glide. Flapping begins to prolong powered flight. From Chatterjee and Templin (2012). Reproduced with permission from Springer Science and Business Media.

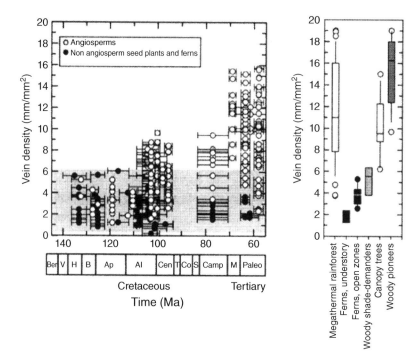

Figure 11.12 Change in angiosperm vein density from the Cretaceous into the Cenozoic (Tertiary). Vein density (D_v) is computed as length of veins in mm per mm² of leaf surface. Data from fossil leaves with ages from 140 to 58 million years old of angiosperms and nonangiosperms are plotted as mean D_v values per morphotype. This shaded area shows the range of D_v for all extinct and living nonangiosperms. Early angiosperms had the same D_v values as nonangiosperms. In the Late Albian (~100 million years ago), the first angiosperm increase in D_v evolved. The second phase of angiosperm D_v increase occurred in the latest Cretaceous (Maastrichtian) and earliest Tertiary (Paleocene) from 68 to 58 million years ago. This is the earliest in Earth history where angiosperms had a similar vein density to the distribution in modern megathermal rain forests, which is shown on the right. This rain forest plot represents 25 species of woody plants from Madang, Papua New Guinea. Horizontal error bars around all data points represent the standard deviation around the mean values for fossil ages. Geological ages: Ber, Berriasian; V, Valanginian; H, Hauterivian; B, Barremian; A, Aptian; Al, Albian; Cen, Cenomanian; T, Turonian; Co, Coniacian; S, Santonian; Camp, Campanian; M, Maastrichtian; Paleo, Paleocene. From Field et al. (2011). Reproduced with permission from the National Academy of Sciences.

Exceptionally preserved *Metasequoia* floodplain and swamp forests are found from Middle Eocene deposits on Axel Heiberg Island in Arctic Canada, where logs, stumps, leaves, and intact treetops are preserved. This variety of paleobotanical information has allowed the structure of these forests to be analyzed. Figure 11.13 shows a three-dimensional reconstruction of these *Metasequoia* forests preserved from two different intervals. These studies demonstrate that in these warm Eocene environments, forests grew above the Arctic Circle and that they were similar to modern North American Pacific Northwest old-growth forests in terms of productivity and biomass.

Summary

From Precambrian microbial communities to Eocene forests of the dawn redwood and from

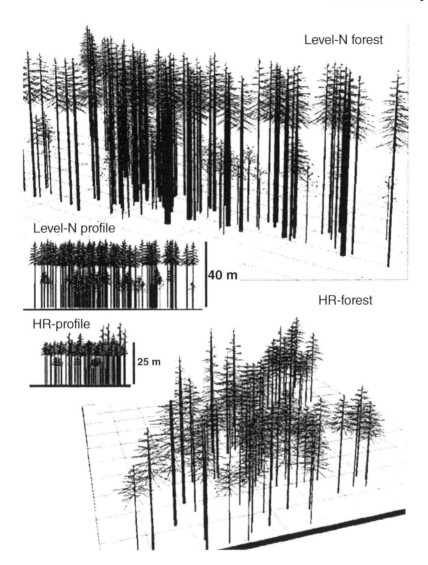

Level-N forest

Level-N profile

40 m

HR-profile

25 m

HR-forest

Figure 11.13 Eocene forests growing above the Arctic Circle. This is a three-dimensional reconstruction of the Napartulik level N and HR fossil forests with vertical profiles, from a study of the high-latitude Middle Eocene deposits of the Buchanan Lake Formation on Axel Heiberg Island, Nunavut Territory, Canada. Spacing and canopy levels are determined from measurements of stumps and trunks and their distributions, together with information from modern *Metasequoia*. From Williams et al. (2003). Reproduced with permission from the Paleontological Society.

Devonian *Tiktaalik* to Jurassic *Archaeopteryx*, the evolutionary paleoecology of terrestrial ecosystems is a fascinating system for study. An integrated paleoecological understanding of these environments is only at its beginning, but new discoveries of fossil deposits and application of new analytical techniques promise much productive future research.

References

Benton, M.J. & Harper, D.A.T. 2009. *Introduction to Paleobiology and the Fossil Record*. Wiley-Blackwell.

Chatterjee, S. & Templin, R.J. 2012. Palaeoecology, Aerodynamics, and the Origin of Avian Flight. *In* Talent, J.A. (ed.), *Earth and Life: Global Biodiversity, Extinction Intervals and Biogeographic Perturbations Through Time*. Springer, pp. 585–612.

Chiappe, L.M. & Dingus, L. 2001. *Walking on Eggs: the Astonishing Discovery of Thousands of Dinosaur Eggs in the Badlands of Patagonia*. Scribner.

DiMichele, W.A., Falcon-Long, H.J., Nelson, W.J., Elrick, S.C. & Ames, P.R. 2007. Ecological gradients within a Pennsylvanian mire forest. *Geology* 35, 415–418.

Eberth, D.A., Robers, R.R. & Fiorillo, A.R. 2007. A Practical Approach to the Study of Bonebeds. *In* Rogers, R.R., Eberth, D.A. & Fiorillo, A.R. (eds.), *Bonebeds: Genesis, Analysis, and Paleobiological Significance*. University of Chicago Press.

Fastovsky, D.E. & Weishampel, D.B. 2009. *Dinosaurs: A Concise Natural History*. Cambridge University Press.

Field, T.S., Brodribb, T.J., Iglesias, A., Chatelet, D.S., Baresch, A., Upchurch, G.R., Jr., Gomez, B., Mohr, B.A.R., Coiffard, C., Kvacek, J. & Jaramillo, C. 2011. Fossil evidence for Cretaceous escalation in angiosperm leaf vein evolution. *Proceedings of the National Academy of Sciences* 108, 8363–8366.

Heers, A.M. & Dial, K.P. 2012. From extant to extinct: locomotor ontogeny and the evolution of avian flight. *Trends in Ecology and Evolution* 27, 296–305.

Ryan, M.J., Russell, A.P., Eberth, D.A. & Currie, P. 2001. The taphonomy of a *Centrosaurus* (Ornithischia: Ceratopsidae) bone bed from the Dinosaur Park Formation (Upper Campanian), Alberta, Canada, with comments on cranial ontogeny. *Palaios* 16, 482–586.

Stanley, S.M. 2008. *Earth System History*. W.H. Freeman.

Williams, C.J., Johnson, A.H., LePage, B.A., Vann, D.R. & Sweda, T. 2003. Reconstruction of Tertiary *Metasequoia* forests. II. Structure, biomass, and productivity of Eocene floodplain forests in the Canadian Arctic. *Paleobiology* 29, 271–292.

Additional reading

Behrensmeyer, A.K., Damuth, J.D., DiMichele, W.A., Potts, R., Sues, H.-D., & Wing, S.L. (eds.). 1992. *Terrestrial Ecosystems Through Time: Evolutionary Paleoecology of Terrestrial Plants and Animals*. University of Chicago Press.

Clack, J.A. 2002. *Gaining Ground*. Indiana University Press.

Currie, P.J., Koppelhus, E.V., Shugar, M.A. & Wright, J.L. (eds.). 2004. *Feathered Dragons*. Indiana University Press.

Gensel, P.G. & Edwards, D. (eds.). 2001. *Plants Invade the Land: Evolutionary & Environmental Perspectives*. Columbia University Press.

Rogers, R.R., Eberth, D.A. & Fiorillo, A.R. (eds.). 2007. *Bonebeds: Genesis, Analysis, and Paleobiological Significance*. University of Chicago Press.

12 Ecological change through time

Introduction

An understanding of change through deep time is one of the unique attributes that paleoecology brings to the table in ecological and evolutionary research. Thus, this theme has already appeared in earlier chapters of this book. For example, ecological trends through the microbially dominated Precambrian into the beginning of the metazoan-dominated Phanerozoic have been discussed in Chapter 7. And, as Chapters 8–11 have detailed, the fossil record becomes abundant with plants and animals in the Phanerozoic, and this is where a variety of trends in these increasingly complex ecosystems have been outlined. This approach of analyzing change in various ecological aspects of the biosphere has been a very fruitful avenue of investigation and represents much of what we know about evolutionary paleoecology – how ecology from various environments on Earth has changed through time and the role that ecological factors play in evolution. Some of the first attempts to understand ecological trends through deep time were on studies of benthic marine tiering (Fig. 1.3) and the three Phanerozoic marine evolutionary faunas (Fig. 3.11). Since then, as discussed in the following text, development of additional trends from marine environments

through the Phanerozoic has been a subject of much research activity.

Diverse approaches for analyzing Phanerozoic trends from marine environments

Many different approaches have been utilized to discern paleoecological trends through deep time and the processes that produced them. For example, taphonomic processes that operate on biotic remains in marine environments have changed with the evolution of new taxa, particularly those with specific behaviors that influence the preservation of fossils (Kidwell and Brenchley 1996). These include diversification through the Phanerozoic of organisms that destroy mineralized skeletons. Such destructive behaviors include bioerosion (Taylor and Wilson 2003) and shell crushing by predators (Figs. 12.1 and 12.13). Similarly, bioturbating organisms, which can mix the temporal relationships of skeletons within sediment, have evolved the ability to burrow to greater depths in soft sediment(Fig. 1.3). These trends indicate an increasingly harsh environment in marine settings for preservation of organic remains before final burial through the Phanerozoic.

Paleoecology: Past, Present and Future, First Edition. David J. Bottjer.
© 2016 John Wiley & Sons, Ltd. Published 2016 by John Wiley & Sons, Ltd.

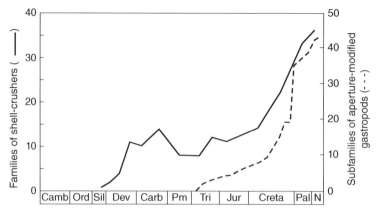

Figure 12.1 Increase through time in diversity of durophagous shell-crushing behavior, and in diversity of gastropods that modify the shell aperture to inhibit predation. Such apertural modifications evolved to inhibit organisms such as crabs which break the margin of the aperture to gain access to the snail within. From McKinney (2007), where additional data sources are indicated. Reproduced with permission from the Columbia University Press.

Many of the trends discussed in Chapter 7 on the ecology of early animals derived from studies of lagerstätten assemblages, such as the Ediacara biota and the Burgess Shale biota. Cambrian communities, when studied from assemblages preserved under normal taphonomic conditions, are typically dominated by trilobites, inarticulate brachiopods, and stem group molluscs (Fig. 12.2). Early work done by Jack Sepkoski and Peter Sheehan on such assemblages within an environmental context showed that during the Cambrian, these typical trilobite-dominated communities of the Cambrian Fauna show a pattern of occurrence in nearshore to slope environments of onshore origination with subsequent restriction to offshore environments. This pattern is illustrated in Fig. 12.3, which also shows that the Paleozoic Fauna began in nearshore environments in the Ordovician and then subsequently migrated to offshore environments. The pattern was repeated again with the evolution of the Modern Fauna in nearshore Ordovician environments (Fig. 12.3), which also migrated offshore through time.

This phenomenon of onshore–offshore change of marine assemblages from the Cambrian to the Ordovician is also reflected by the increasing diversity of animals that had evolved carbonate skeletons through this time (Fig. 12.4). In particular, the number of genera of different calcifying

organisms increases dramatically from the Cambrian into the Ordovician in what is known as the great Ordovician biodiversification event (GOBE), so that the fossil record becomes significantly richer in carbonate bioclasts with the evolution of the Paleozoic Fauna.

This change in type as well as increase in abundance of different types of bioclasts is reflected in the development of shell beds, including their taxonomic composition and thickness. For example, a study of shell beds shown in Fig. 12.5 indicates that at the beginning of the Ordovician, shell beds were dominated by trilobites and echinoderms but in younger strata of the Middle Ordovician, shell beds in this area have fewer trilobites, and greater and greater numbers of other taxa such as brachiopods. These changes in shell bed composition are a good estimate of the changes in abundance of organisms as shallow marine ecosystems shifted from the trilobite-dominated Cambrian Fauna to the brachiopod-dominated Paleozoic Fauna.

Part of the diversification of the GOBE and the proliferation of the Paleozoic Fauna was due to the increased utilization of ecospace represented by an increase in epifaunal tiering from the Cambrian into the Ordovician (Fig. 1.3). This increased use of epifaunal ecospace was maintained through the Paleozoic by organisms such as bryozoans and a variety of echinoderms, particularly crinoids

(A)

(B)

Figure 12.2 Shelf-depth Cambrian benthic communities from fossil assemblages preserved under normal taphonomic conditions. (A) Shallow-water shelly community consisting of trilobites (a, b, c, i), inarticulate brachiopods (d, f), articulate brachiopods (e), hyoliths (g), and tubular remains of uncertain affinities (h).

(B) Deeper shelf shelly community of trilobites (a, d, e, f, i), inarticulate brachiopods (c), hyoliths (g), and monoplacophorans (h). From McKinney (2007). Reproduced with permission from the Columbia Univeristy Press.

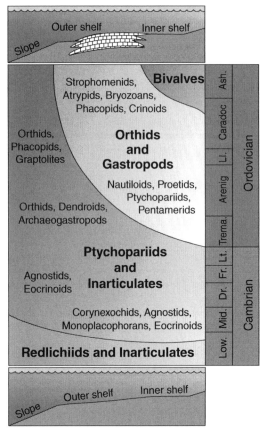

Figure 12.3 Onshore–offshore occurrence of assemblages during the Cambrian and Ordovician. Taxa in larger bold lettering are those which were most significant in differentiating the assemblages through cluster analysis (Sepkoski and Sheehan 1983). The lower two groups of assemblages correspond to the Cambrian marine evolutionary fauna, the third group of assemblages corresponds to the Paleozoic marine evolutionary fauna, and the assemblages dominated by bivalves correspond to the Modern marine evolutionary fauna (see Fig. 3.11). Trilobites include redlichiids, corynexochids, agnostids, ptychopariids, phacopids, and proetids; brachiopods include inarticulates, orthids, atrypids, and strophomenids. Below, the classic onshore–offshore environmental gradient; above, the same gradient modified with the addition of midshelf carbonate buildups. From Harper (2006). Reproduced with permission from Elsevier. (*See insert for color representation.*)

(Fig. 12.6). Infaunal tiering by burrowing organisms increased significantly in the later Paleozoic (Fig. 1.3). Although much of the Mesozoic tiering was well developed in epifaunal and infaunal ecospace, epifaunal tiering began to be significantly reduced toward the end of the Mesozoic (Fig. 1.3). This trend in community composition through the Mesozoic and Cenozoic of increased infaunalization (Fig. 12.7) is largely due to increased dominance by the Modern Fauna.

Counts of abundance to determine the ecologically dominant organisms in paleocommunities are another important component of analyzing ancient ecology. Figure 12.8 shows an analysis of trends in abundance and distribution in assemblages from the Ordovician into the Cenozoic. These trends highlight changes in dominance in marine benthic paleocommunities through this broad Phanerozoic time interval. The trends illustrate how Paleozoic communities dominated by rhynchonelliform brachiopods declined in abundance whereas bivalves, a characteristic component of the Modern Fauna, increased significantly, as did gastropods. During this time interval, the components of the Modern Fauna became not only more diverse but also more abundant and dominant in ancient communities.

Further analysis of shell beds through the Phanerozoic shows that they are typically thinner in the Paleozoic as compared to the Mesozoic and Cenozoic (Fig. 12.9). This trend is largely due to the replacement of the Paleozoic Fauna by the Modern Fauna in benthic environments. Much of the transition between these two evolutionary faunas occurred because the end-Permian mass extinction preferentially affected organisms of the Paleozoic Fauna (Figs. 3.11, 3.13, and 12.8). A detailed study of shell beds has been used to understand geographic differences in the transition from the Paleozoic to the Modern Fauna in the early Mesozoic after the end-Permian mass extinction. Figure 12.10 illustrates a compositional analysis of Middle Triassic shell beds from several regions around the world. In British Columbia, they are dominated by components of the Paleozoic Fauna such as

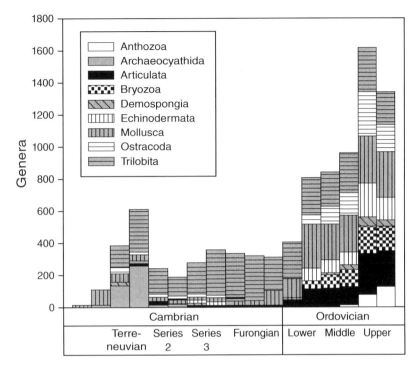

Figure 12.4 Increase in diversity of marine organisms with carbonate skeletons from the Cambrian into the Ordovician. This plot of generic diversity of calcifying animal groups shows the increase that occurred due to the evolution of the Paleozoic Fauna as part of the GOBE. From Pruss et al. (2010). Reproduced with permission from the SEPM Society for Sedimentary Geology.

brachiopods and crinoids; while in Germany, there is a mix of shell beds dominated by the Modern and Paleozoic Faunas. Thus, although the diversity of these Paleozoic Fauna components is not high in the Middle Triassic, in certain settings, they still could act as abundant and dominant organisms in these communities. This trend continued into the Jurassic (Fig. 12.9).

The trend in increased ecospace utilization with evolution from the Cambrian to the Paleozoic to the Modern Fauna is reflected in the classic analysis by Bambach on how ecological megaguilds filled up through the Phanerozoic (Fig. 12.11). With increased diversity of occupied "Bambachian megaguilds," there was also greater diversity of organisms in each of these megaguilds through time (Fig. 12.11). Using the Bush cube (Fig. 3.14), guild structure can be understood at a finer level of detail.

Figure 12.12 illustrates a study using the Bush cube to analyze ecospace utilization from the Ediacaran and Cambrian to the Recent which highlights the dramatic difference between ecospace utilization by marine animals before and during the Cambrian explosion with the extensive ecospace utilization found in modern communities.

An ecological aspect of marine environments that has a strong influence on ecospace utilization is predation. In the Mesozoic, an increase in predators (Fig. 12.13) in benthic marine environments affected ecospace occupation and evolution of many invertebrate animals. This phenomenon, as outlined by Geerat Vermeij and known as the Mesozoic Marine Revolution, led to an increase of infaunalization, as a means toward escaping seafloor predators, and a loss of epifaunal tiering (Fig. 1.3), as these exposed organisms, many of which were

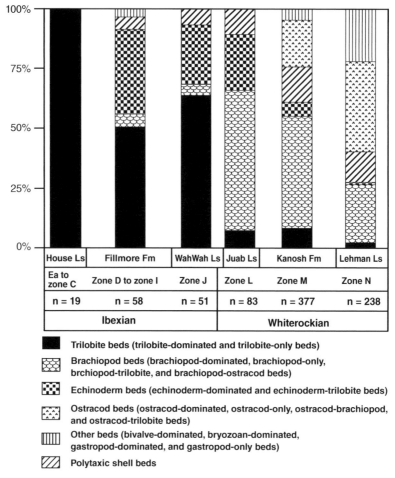

Figure 12.5 Distribution of taxonomic types of shell beds from Basin and Range (United States) Lower Ordovician (Ibexian) and Middle Ordovician (Whiterockian) strata. Data used are primarily from stratigraphic units in the Ibex area of western Utah and also from their correlatives in eastern Nevada. *N* is the number of shell beds examined in the corresponding stratigraphic units. Total number of shell beds examined is 826. From Li and Droser (1999). Reproduced with permission from the SEPM Society for Sedimentary Geology.

fixed to the seafloor, were subject to more intense predation. Detailed studies of predation have shown an increase in features such as drill holes and repair scars, considered to be a record of durophagous predation, since the mid-Cretaceous (Fig. 12.13). Adaptations to this increased level of durophagous predation, such as the increase in gastropod apertural modifications (Fig. 12.1) or spinosity of many epifaunal invertebrates (Fig. 12.13), appear to have also increased during this time. Several other modes of predation, such as shell crushing or breaking by demersal vertebrates, show an increase in the Late Triassic (Fig. 12.13). The Late Triassic also saw an increase in adaptations by prey to avoid predation, such as cementing by bivalves to a substrate (Fig. 12.13), which decreases the ability of a predator to manipulate these as prey items, and hence to proceed to successful predation. Thus,

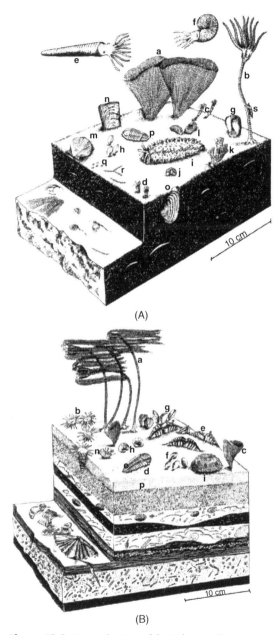

(A)

(B)

Figure 12.6 Diversification of the Paleozoic Fauna in shelf-depth environments. (A) Carboniferous mud-dwelling community consisting of stenolaemate bryozoans (a, k, r), crinoids (b), corals (c, d), cephalopods (e, f), articulate brachiopods (g, h, j, l), holothurians (i), bivalves (m, n, o), trilobites (p), ostracodes (q), and gastropods (s). (B) Devonian mud-dwelling community consisting of crinoids (a), corals (b, i), stenolaemate bryozoans (c), trilobites (d), and articulate brachiopods (e, f, g, h). From McKinney (2007). Reproduced with permission from Columbia University Press.

the Late Triassic may have been the time when the Mesozoic Marine Revolution was initiated (Fig. 12.13).

Onshore–offshore patterns, such as those examined for the Cambrian and Paleozoic Faunas in the Paleozoic, can also be studied through analysis of individual clades. This has been done for a number of taxa in the post-Paleozoic, where a variety of patterns have been discerned. As for the Cambrian and Paleozoic Faunas, some post-Paleozoic clades first evolved in nearshore environments and then progressively occupied offshore environments and at some point were then restricted to these offshore environments. This onshore–offshore pattern is exhibited by the isocrinid crinoids, a major group of stalked articulate crinoids, which are epifaunal suspension feeders and considered to be part of the Paleozoic Fauna (Fig. 3.11). These stalked crinoids first evolved in nearshore environments and by the Jurassic were living across the shelf and into deep basins (Fig. 12.14). However, by the Late Cretaceous, they began to disappear from nearshore environments so that they then were only found in outer shelf, slope, and deep basin settings (Fig. 12.14). It is thought that this onshore–offshore evolutionary pattern occurred because a predator, probably some type of fish, evolved in shelf environments in the Late Cretaceous, and because isocrinid crinoids were unable to adapt to avoid this predation, they now are only found in deeper water where this predator does not live. The retreat of stalked crinoids to deep water is one of the main reasons why epifaunal tiering showed a significant decrease in the Cretaceous (Fig. 1.3). Thus, the evolutionary and ecological effects of increased predation, as exemplified by the Mesozoic Marine Revolution, played a significant role in completing the transition to dominance by the Modern Fauna in marine paleocommunities during the Cretaceous.

Similar analyses of other post-Paleozoic clades show a variety of different patterns. Figure 12.15 shows a plot on a time–environment diagram of the number of genera of cheilostome bryozoans within paleocommunities. Cheilostome bryozoans, which are epifaunal suspension feeders, first evolved in the Late Jurassic and are considered to be part of the Modern Fauna (Fig. 3.11). For a long time, cheilostomes were rare, with only one genus present in any given paleocommunity (Fig. 12.15). But

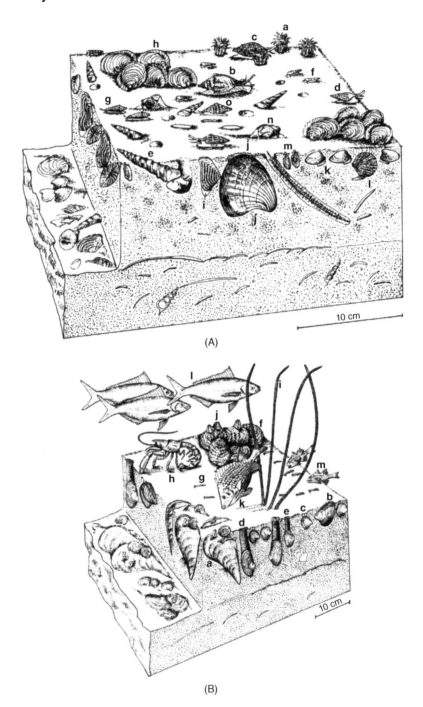

Figure 12.7 Shelf-depth Paleogene benthic communities. (A) Eocene silt/sand community consisting of corals (a), gastropods (b, c, d, e, f, g, n, o), bivalves (h, I, j, k, l, m), and polychaetes (occupying burrow). (B) Eocene sandy clay community consisting of bivalves (a, b, c, d, e, f), annelids (g), malacostracans (h), octocorals (i), barnacles (j), and osteichthyes (k, l, m). From McKinney (2007). Reproduced with permission from Columbia University Press.

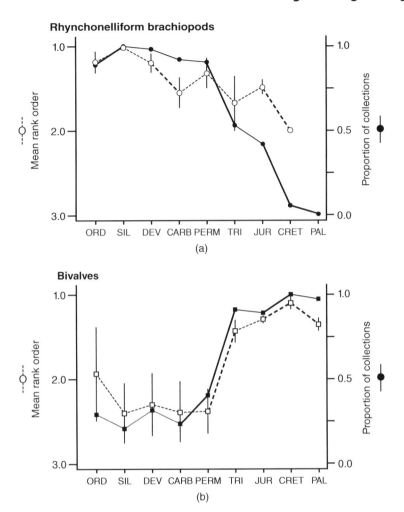

Figure 12.8 Trends in ecological dominance in benthic assemblages through time. Ecological dominance can be assessed by computing mean rank order for sampled assemblages, with rank order being the rank in abundance that a taxon has within an assemblage or paleocommunity; the most dominant taxon has a rank order of 1. Breadth of distribution, another measure of ecological dominance, is computed as the percent of studied assemblages in which the taxon is present. These two ecological dominance metrics are plotted for (a) rhynchonelliform brachiopods, (b) bivalves, and (c) gastropods. Trends in mean rank-order abundance are indicated by dashed lines and open symbols, and breadth of distribution is indicated by solid lines and filled symbols. The scale has been inverted so that higher areas of the curve correspond to increasing abundance (and smaller rank orders). Error bars indicate 95% confidence intervals, and significant period-to-period changes (indicated by a Mann–Whitney U-test for rank order and a Z-test for breadth of distribution) are shown by thick black lines joining successive values. From Clapham et al. (2006). Reproduced with permission from the SEPM Society for Sedimentary Geology.

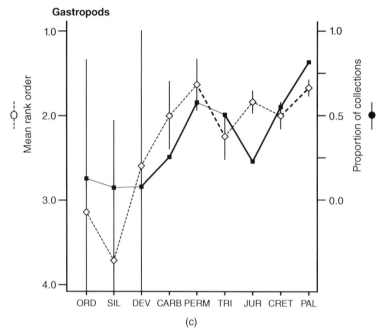

Figure 12.8 (*continued*)

by the mid-Cretaceous, they evolved rapidly so that in the Tertiary there were as many as 30 or more genera in individual shelf communities (Fig. 12.15). Cheilostome bryozoans showed a nearshore origination and offshore migration before their explosive diversification in the Late Cretaceous and became more diverse in offshore than onshore paleocommunities in the Tertiary (Fig. 12.15).

The same sort of analysis that was used for bryozoans has also been used to examine the paleoenvironmental history of tellinacean bivalves. This common infaunal bivalve group, which includes both suspension and deposit feeders and is part of the Modern Fauna (Fig. 3.11), does not show a typical onshore–offshore pattern but rather evolved to live in inner and middle shelf environments from the Triassic onward (Fig. 12.16).

The varying paleoenvironmental histories that can be constructed for these three post-Paleozoic clades indicates that onshore–offshore patterns are not due to a relatively coherent community evolving in nearshore environments and then moving offshore. Rather, they are the product of individual clade paleoenvironmental histories which favor onshore to offshore outcomes through time. Overall, the record of occurrence of innovations in onshore environments may have been caused by the greater likelihood for evolutionary innovations to happen in onshore rather than offshore settings or that innovations preferentially survive in onshore settings.

Marine hard substrate nonreef communities include those on rocks, wood, and sedimentary hardgrounds and are composed of encrusters and organisms that bore into hard substrates. The increase of bioerosion by boring organisms in these paleocommunities through the Phanerozoic (Taylor and Wilson 2003) may be analogous to the increase of infaunal tiering that occurs in Phanerozoic soft substrates (Fig. 1.3). Encrusting communities may also show exposure to increased predation from the Paleozoic into the Mesozoic and Cenozoic, demonstrated as an increase in encruster skeletalization through this time (Taylor and Wilson 2003). These trends show increased ecospace utilization through time overlain with response to predation exemplified by the effects of the Mesozoic Marine Revolution.

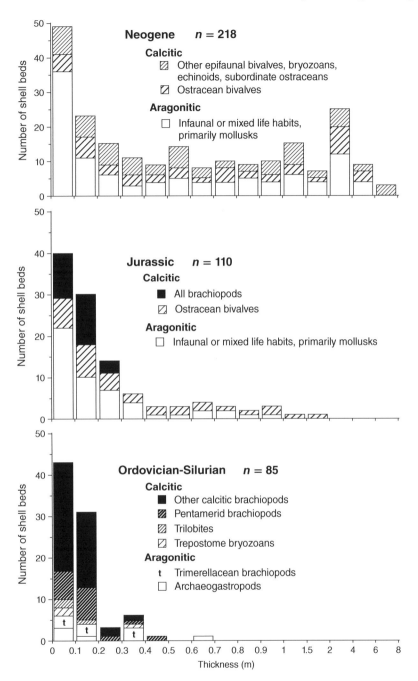

Figure 12.9 Shell bed thickness trends from the Paleozoic into the Cenozoic. This illustrates frequency distributions of thickness and composition of bioclast-supported shell beds (bioclasts >2 mm, excluding crinoids, and hence encrinites), showing a significant shift in overall shell bed thickness to right-skewed distributions over time. Where thicknesses fell at a size boundary, shell beds were tallied in the larger size class. The contrast between Ordovician–Silurian and Neogene data sets reflects the evolution of several bioclast-producing groups and does not depend upon a single biomineral or life habit. From Kidwell and Brenchley (1994). Reproduced with permission from the Geological Society of America.

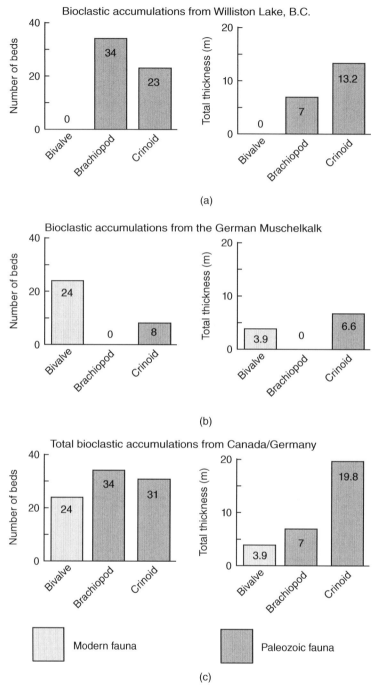

Figure 12.10 Middle Triassic shell beds and the transition from the Paleozoic to the Modern Fauna. Number of shell beds comprised of each bioclast producer (left), and combined thickness of shell beds comprised of each bioclast producer (right). (a) Paleozoic faunal constituents (brachiopods and crinoids) are the primary shell bed producers from Williston Lake, British Columbia. (b) More shell beds are produced by Modern Fauna (bivalves) than Paleozoic Fauna (crinoids) from the German Muschelkalk; however, the total thickness of encrinites exceeds that of bivalve shell beds. (c) Pooled data from both Canada and Germany, showing that the majority of shell beds are dominated by Paleozoic Fauna, with crinoids producing the greatest combined total thickness. From Greene et al. (2011). Reproduced with permission from Elsevier.

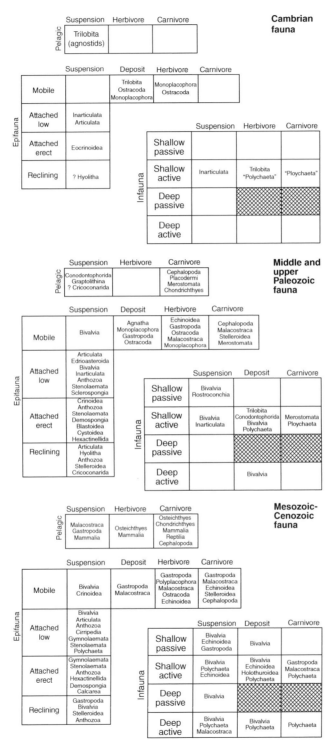

Figure 12.11 Increased ecospace utilization through the Phanerozoic. These charts show the adaptive strategies of taxa that comprised biotas during the Cambrian, the Middle and Upper Paleozoic, and the post-Paleozoic. Each adaptive strategy, or Bambachian megaguild, is represented by its own box and is defined by relationship to substrate and feeding mode. Deep passive herbivore, carnivore, and deposit-feeding modes are not possible. From Foote and Miller (2006). Reproduced with permission from W.H. Freeman.

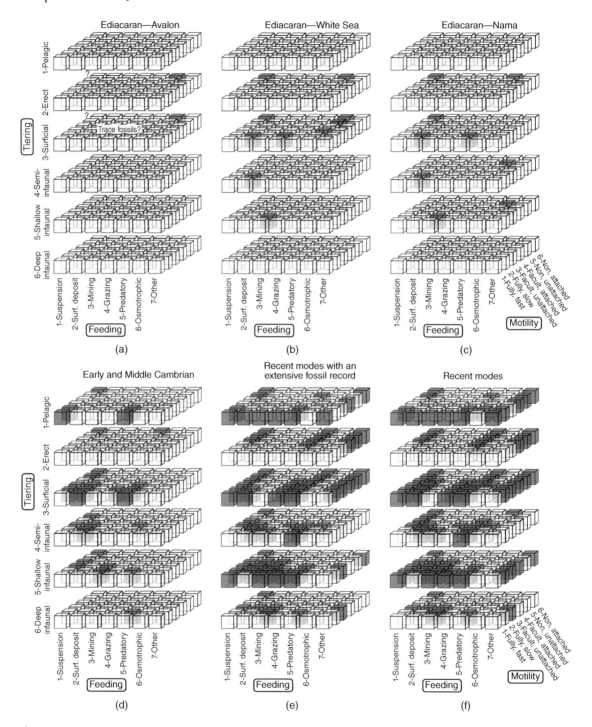

Figure 12.12 Ecological utilization in early animal faunas and the modern marine fauna using the Bush cube (Fig. 3.13). Black boxes denote modes of life occupied by at least one species in a fauna, and white boxes denote unoccupied modes. Gray boxes and question marks in (a) indicate uncertainty. (a) The Avalon assemblage of the Ediacaran Period, (b) the White Sea assemblage of the Ediacaran Period, (c) the Nama assemblage of the Ediacaran Period, (d) the early–middle Cambrian Period, (e) Recent animals that have an extensive fossil record, and (f) all Recent animals. From Bush et al. (2011). Reproduced with permission from Springer Science and Business Media. (*See insert for color representation.*)

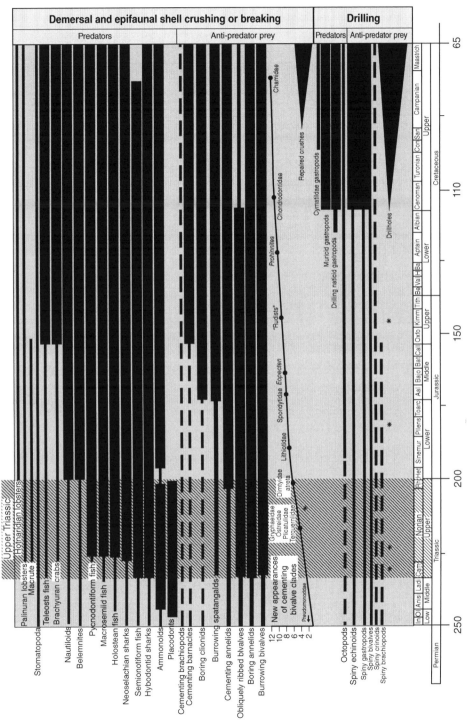

Figure 12.13 Components of the Mesozoic Marine Revolution. This timescale shows various predator taxa, prey taxa with antipredator adaptations, and other key events in the Mesozoic. The thick black lines indicate diversifications and greater abundances, thin horizontal lines are taxa that are present but not abundant, and dashed lines are groups that were likely to be present but have sparse fossils. From Tackett and Bottjer (2012). Reproduced with permission from the SEPM Society for Sedimentary Geology.

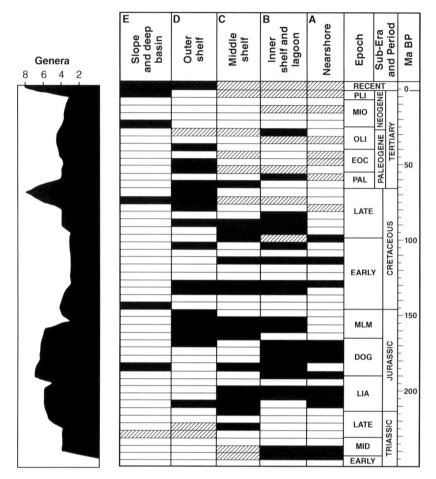

Figure 12.14 Onshore–offshore evolutionary history for isocrinid crinoids. This time–environment diagram for isocrinid crinoids shows presence–absence (right), accompanied by a plot of isocrinid global generic diversity (left). Black boxes indicate presence of isocrinids in an assemblage of that age and environment; boxes with slash pattern indicate presence of other echinoderm fossils, which constitute the taphonomic control group for this study. A taphonomic control group is an organism(s) that produces fossil remains that have a similar taphonomic behavior to the studied organism, thus indicating that the absence of the studied organisms is not due to taphonomic conditions. Paleoenvironmental categories are defined in Bottjer and Jablonski (1988). From Bottjer and Jablonski (1988). Reproduced with permission from the SEPM Society for Sedimentary Geology.

The study of broad trends through time for reefs and other buildups has typically been done separately from studies of such trends in level-bottom settings. To some large extent, this approach has been taken because reefs behave somewhat differently from level-bottom communities. In particular, reefs appear to be more sensitive to environmental crises than paleocommunities in level-bottom settings and thus are viewed as "the canary in the coal mine" for marine environments and reactions to increased environmental stress. This topic has been addressed in Chapter 9 through examination of specific examples, but more detailed patterns are outlined in Fig. 12.17. The biotic effects

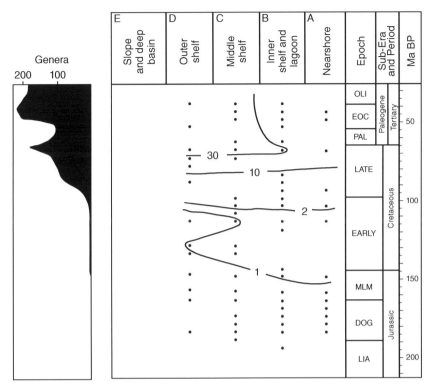

Figure 12.15 Paleoenvironmental history for cheilostome bryozoans. This time–environment diagram is contoured for cheilostome bryozoan generic richness (right), accompanied by a plot of Jurassic–Paleogene cheilostome global generic diversity (left). Dots indicate location of data points, each of which is generic richness for a single assemblage. Paleoenvironmental categories defined in Bottjer and Jablonski (1988). From Bottjer and Jablonski (1988). Reproduced with permission from the SEPM Society for Sedimentary Geology.

of mass extinctions punctuate the reef record to a greater extent than they typically do in level-bottom environments. Mass extinctions had a particularly devastating effect upon archaeocyaths in the early Cambrian, stromatoporoids in the Late Devonian, rugose and tabulate corals at the end of the Permian, and rudist bivalves at the end of the Cretaceous (Fig. 12.17). These effects of mass extinction overlie the broad pattern of microbial reefs and mounds dominating through the Precambrian, with their reduction into the Phanerozoic, followed by evolution of various reef-building animals (Fig. 12.17). Corals were not always the dominant organisms in Phanerozoic reefs, and sponges,

including stromatoporoids, as well as bivalves also have played a large role (Fig. 12.17).

In the oceans, the environmental distribution of organisms that make calcium carbonate skeletons has a strong effect on biogeochemical cycles. The environments inhabited by the Cambrian, Paleozoic, and Modern Faunas, all of which primarily produce calcium carbonate skeletons, are largely nearshore and shelf settings. Until the Jurassic, most carbonate skeletons deposited in the oceans were from components of these three evolutionary faunas and hence occurred on shelf seafloors, so that the areal distribution of seafloor calcium carbonate sediments was relatively restricted. However, the evolution at the end of the Triassic and beginning

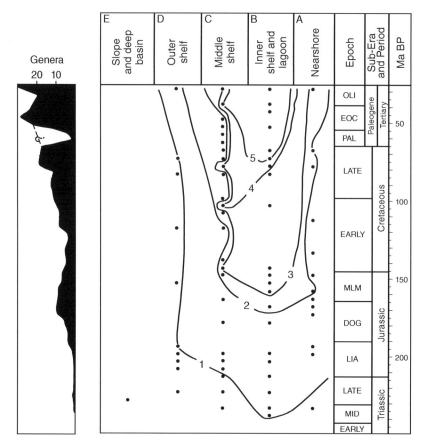

Figure 12.16 Paleoenvironmental history for tellinacean bivalves. This time–environment diagram is contoured for generic richness (right), accompanied by a plot of tellinacean Triassic–Paleogene global generic diversity (left). Dots indicate location of data points, each of which is generic richness for a single assemblage. Paleoenvironmental categories defined in Bottjer and Jablonski (1988). From Bottjer and Jablonski (1988). Reproduced with permission from the SEPM Society for Sedimentary Geology.

of the Jurassic of planktonic foraminifera and coccolithophores, which have calcium carbonate skeletons, caused significant areas of the deep seafloor also to be covered in calcium carbonate sediment, greatly increasing the ability of the ocean to buffer changes in ocean carbonate chemistry. Thus, before the Jurassic, the oceans had a relatively reduced buffering capacity, and this condition is known as the Neritan ocean (Fig. 12.18). The increased buffering capacity represented by the increase in seafloor covered in carbonate skeletons,

beginning in the Jurassic, is known as the Cretan ocean (Fig. 12.18). As will be discussed in the final two chapters, the mass extinctions at the end of the Permian and the end of the Triassic were likely ultimately due to enormous eruptions of basalt, called large igneous provinces, which would have injected a significant amount of CO_2 into the atmosphere and affected ocean carbonate chemistry. It has been hypothesized that eruptions of large igneous provinces when Earth had a Neritan ocean would have caused greater mass extinctions than

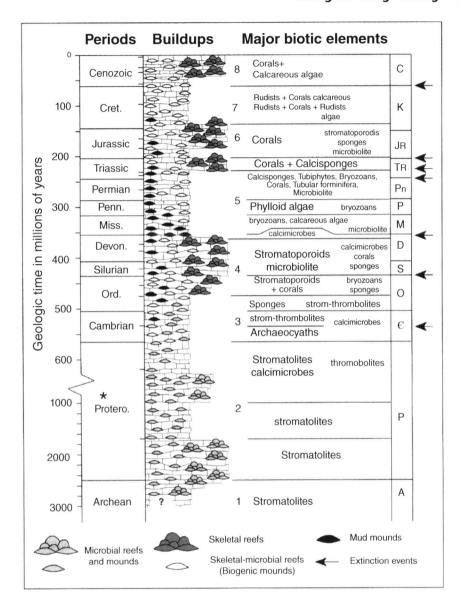

Figure 12.17 History of buildups through geological time. This is an idealized stratigraphic column representing geologic time and illustrating periods when there were only microbial reefs and mounds (indicated by light gray symbols below the Cambrian) from periods with skeletal reefs (indicated by dark gray symbols above the Cambrian) and skeletal–microbial reefs (biogenic mounds) (indicated by white symbols above the Proterozoic) as well as mud mounds (indicated by black symbols above the Proterozoic). Numbers indicate different associations of reef- and mound-building biota. Arrows signal major extinction events; * indicates scale change. From James and Wood (2010). Facies Models. Reproduced with permission from Geological Association of Canada. (*See insert for color representation.*)

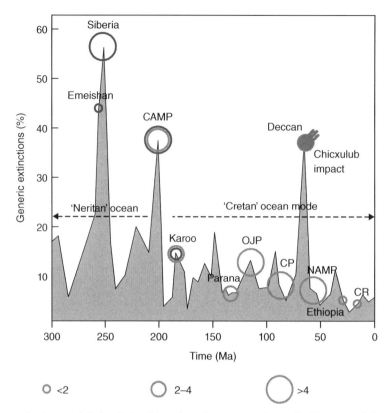

Figure 12.18 Mass extinctions and their relationship to large igneous provinces, the Neritan and Cretan oceans, and the Chicxulub impact. Mass extinctions are indicated by significant increases in generic extinction rates. Occurrences of large igneous provinces are indicated by circles overlying the extinction rate curve; the scale of circle sizes for large igneous provinces is in millions of cubic kilometers of basalt. Siberia, Siberian Traps; CAMP, Central Atlantic Magmatic Province; Deccan, Deccan traps; NAMP, North Atlantic Magmatic Province; OJP, Ontong Java; CP, Caribbean Plateaux; CR, Columbian River basalts. The end-Permian mass extinction occurred at the same time as the eruption of the Siberian traps, the end-Triassic mass extinction occurred at the same time as the eruption of CAMP, and the end-Cretaceous mass extinction is associated with the eruption of the Deccan traps and the Chicxulub impact. Note that eruption of large igneous provinces when there is a Cretan ocean did not lead to increases in extinction rate as compared to the Siberian trap and CAMP eruptions when a Neritan ocean existed, other than for the Deccan traps and the end-Cretaceous mass extinction, which may have been largely caused by the Chicxulub impact. From Sobolev et al. (2011). Reproduced with permission from Macmillan Publishers Ltd. (*See insert for color representation.*)

large igneous province eruptions when there is a Cretan ocean (Fig. 12.18).

The Mesozoic evolution of coccolithophorids and other red phytoplankton such as dinoflagellates and diatoms (Fig. 12.19) is thought to have had other significant effects on the ocean biosphere. These primary producers have a greater nutrient content than the green phytoplankton which were prevalent in the Paleozoic (Fig. 12.19). Strontium isotope ratios indicate that this evolutionary burst of red phytoplankton in the Mesozoic and Cenozoic was because of increased nutrient delivery to ocean ecosystems (Fig. 12.19). This increase in nutrient availability and abundance and diversity of red phytoplankton is likely also one of the factors that led to the diversification of the Modern Fauna (Fig. 12.19).

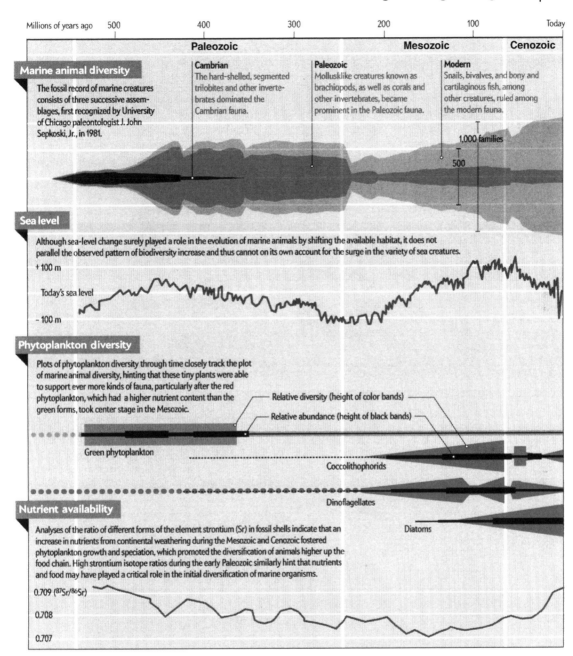

Figure 12.19 Mesozoic evolution of phytoplankton and the Modern Fauna. A mass extinction at the end of the Paleozoic, around 252 million years ago, preferentially affected the Paleozoic Fauna leading to the takeover by the Modern Fauna. The diversification of the Modern Fauna has been attributed to a variety of biotic factors as well as physical factors such as sea-level change. However, Mesozoic and Cenozoic increased nutrient flux into the oceans may have spurred the evolution of phytoplankton, the tiny plants that form the base of marine food chains, leading to increased diversification of the Modern Fauna. From Martin and Quigg (2013), where data sources are indicated. (*See insert for color representation.*)

Summary

These ecological trends through time illustrate the dynamic nature of evolutionary paleoecology. A challenge and opportunity for the future will be to further develop our understanding of ecological trends through time in terrestrial environments (briefly outlined in Chapter 11), and to integrate these with marine trends, for a holistic understanding of the evolutionary paleoecology of life on Earth.

References

Bottjer, D.J. & Jablonski, D. 1988. Paleoenvironmental patterns in the evolution of post-Paleozoic benthic marine invertebrates. *Palaios* 3, 540–560.

Bush, A.M., Bambach, R.K. & Erwin, D.H. 2011. Ecospace Utilization During the Ediacaran Radiation and the Cambrian Eco-explosion. *In* Laflamme, M., Schiffbauer, J.D. & Dornbos, S.Q. (eds.), *Quantifying the Evolution of Early Life: Numerical Approaches to the Evaluation of Fossils and Ancient Ecosystems. Topics in Geobiology 36.* Springer Science, pp. 111–134.

Clapham, M.E., Bottjer, D.J., Powers, C.M., Bonuso, N., Fraiser, M.L., Marenco, P.J., Dornbos, S.Q. & Pruss, S.B. 2006. Assessing the ecological dominance of Phanerozoic marine invertebrates. *Palaios* 21, 431–441.

Foote, M. & Miller, A.I. 2006. *Principles of Paleontology,* 3rd *Edition*. W.H. Freeman and Company.

Greene, S.E., Bottjer, D.J., Hagdorn, H. & Zonneveld, J.-P. 2011. The Mesozoic return of Paleozoic faunal constituents: A decoupling of taxonomic and ecological dominance during the recovery from the end-Permian mass extinction. *Palaeogeography, Palaeoclimatology, Palaeoecology* 308, 224–232.

Harper, D.A.T. 2006. The Ordovician biodiversification: Setting an agenda for marine life. *Palaeogeography, Palaeoclimatology, Palaeoecology* 212, 148–166.

James, N.P. & Wood, R. 2010. Reefs. *In* James, N.P. & Dalrymple, R.W., *Facies Models 4.* Geological Association of Canada, pp. 421–447.

Kidwell, S.M. & Brenchley, P.J. 1994. Patterns in bioclastic accumulation through the Phanerozoic: Changes in input or in destruction? *Geology* 22, 1139–1143.

Kidwell, S.M. & Brenchley, P.J. 1996. Evolution of the Fossil Record: Thickness Trends in Marine Skeletal Accumulations and Their Implications. *In* Jablonski, D., Erwin, D.H. & Lipps, J.H. (eds.). *Evolutionary Paleobiology*. University of Chicago Press, pp. 290–336.

Li, X. & Droser, M.L. 1999. Lower and Middle Ordovician shell beds from the Basin and Range Province of the western United States (California, Nevada, and Utah). *Palaios* 14, 215–233.

Martin, M. & Quigg, A. 2013. Tiny plants that once ruled the seas. *Scientific American* 308, 40–45.

McKinney, F.K. 2007. *The Northern Adriatic Ecosystem.* Columbia University Press.

Pruss, S.B., Finnegan, S., Fischer, W.W. & Knoll, A.H. 2010. Carbonates in skeleton-poor seas: New insights from Cambrian and Ordovician strata of Laurentia. *Palaios* 25, 73–84.

Sepkoski, J.J., Jr. & Sheehan, P.M. 1983. Diversification, Faunal Change, and Community Replacement During the Ordovician Radiations. *In* Tevesz, M.J.S. & McCall, P.L. (eds.), *Biotic Interactions in Recent and Fossil Benthic Communities. Topics in Geobiology* 3. Plenum Press, pp. 673–718.

Sobolev, S.V., Sobolev, A.V., Kuzmin, D.V., Krivolutskaya, N.A., Petrunin, A.G., Arndt, N.T., Radko, V.A. & Vasiliev, Y.R. 2011. Linking mantle plumes, large igneous provinces and environmental catastrophes. *Nature* 477, 312–316.

Tackett, L.S. & Bottjer, D.J. 2012. Faunal succession of Norian (Late Triassic) level-bottom benthos in the Lombardian Basin: Implications for the timing, rate, and nature of the early Mesozoic marine revolution. *Palaios* 27, 585–593.

Taylor, P.D. & Wilson, M.A. 2003. Palaeoecology and evolution of marine hard substrate communities. *Earth-Science Reviews* 62, 1–103.

Additional reading

Bush, A.M., Pruss, S.B., & Payne, J.L. (eds.). 2013. Ecosystem Paleobiology and Geobiology. *The Paleontological Society Papers* 19. The Paleontological Society.

Kelley, P.H., Kowalewski, M. & Hansen, T.A. (eds.). 2003. Predator–Prey Interactions in the Fossil Record. *Topics in Geobiology* 20. Kluwer Academic/Plenum Publishers.

Tevesz, M.J.S. & McCall, P.L. (eds.). 1983. *Biotic Interactions in Recent and Fossil Benthic Communities. Topics in Geobiology* 3. Plenum Press.

13 Ecological consequences of mass extinctions

Introduction

Mass extinctions are well known as biodiversity crises that affect wide geographic areas and affect a broad taxonomic spectrum in which extinction rates increase significantly from normal background rates. But mass extinctions are also ecological crises. The most prominent mass extinctions in the Phanerozoic are the "Big Five," at the end of the Ordovician, in the Late Devonian, at the end of the Permian, at the end of the Triassic, and at the end of the Cretaceous (Fig. 13.4). The most well-known mass extinction is at the end of the Cretaceous, when the dinosaurs became extinct. This mass extinction is generally thought to have been primarily caused by impact of an extraterrestrial object, most likely an asteroid (Fig. 12.18). Other mass extinctions, such as those at the end of the Permian and the Triassic, are thought to have been due to the multiple effects of emplacement of large igneous provinces (LIPs) (Fig. 12.18), including injection of greenhouse gases into the atmosphere and subsequent global warming, oceanic anoxia, and oceanic acidification.

End-Permian mass extinction

Perhaps the most severe biotic crisis in Earth history was the end-Permian mass extinction (~252.6 million years ago), with biotic recovery not occurring until almost 5 million years later at the start of the Middle Triassic (Anisian). The interval marked by the end-Permian mass extinction and the subsequent Early Triassic experienced significantly increased warming due to eruption of the Siberian Traps LIP. Baking and burning of coal and other organically enriched deposits during intrusion and eruption of this LIP (Figs. 13.1 and 13.11) enhanced the volcanic contribution of greenhouse gases to the atmosphere, particularly carbon dioxide. Studies of carbon isotopes from the Nanpanjiang Basin of southern China from the end of the Permian to the Middle Triassic have produced a trend which is thought to represent a global signal and shows significant volatility during the Early Triassic, starting with a negative excursion at the Permian–Triassic boundary (Fig. 13.2). Continued negative excursions through the Early Triassic (Fig. 13.2) have been interpreted as reflecting periodic eruptions of the Siberian Traps during this time, which disrupted the carbon cycle. The carbon isotope record has served not only as an indicator of environmental and biotic changes occurring during the Early Triassic but also as a correlation tool in a variety of studies. For example, more recent studies from the Nanpanjiang Basin utilizing carbon isotopes of carbonates as well as oxygen isotopes from conodont phosphatic skeletal elements have shown that Earth became very hot during this time (Fig. 13.3), due to the

Paleoecology: Past, Present and Future, First Edition. David J. Bottjer.
© 2016 John Wiley & Sons, Ltd. Published 2016 by John Wiley & Sons, Ltd.

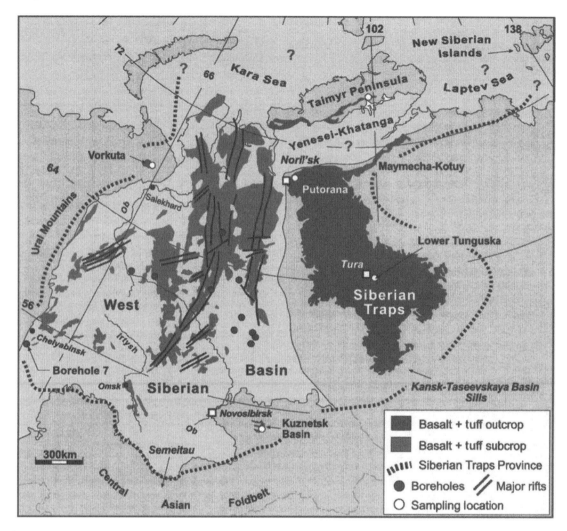

Figure 13.1 Extent of the Siberian traps flood basalt province (Russia), showing basalt and tuff in outcrop and the subsurface. This Siberian volcanism, which occurred for several million years beginning at the end of the Permian, may have extended over ~5 million km², with a volume of 3–5 milion km³. The Paleozoic sedimentary rocks through which these volcanic rocks were intruded and erupted include extensive carbonate and coal deposits. From Reichow et al. (2009), where information on studied boreholes and sampling locations, as well as other data sources, can be found. Reproduced with permission from Elsevier.

injection of carbon dioxide and other greenhouse gases caused by Siberian Trap eruptions.

The development of microbial structures in marine carbonate rocks (Fig. 13.4), as well as wrinkle structures in siliciclastic settings (Fig. 13.5), was unusually common during the Early Triassic, implying an ocean more dominated by microbes than is typical for most of the Phanerozoic. Correlation of the occurrence of stromatolites, wrinkle structures, and other anachronistic facies in the Early Triassic with negative carbon isotope excursions (Fig. 13.6) has suggested a potential link

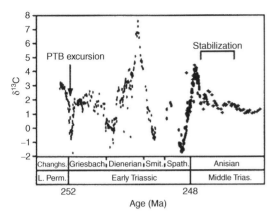

Figure 13.2 Composite carbon isotope record from the Upper Permian through Middle Triassic. This is compiled from stratigraphic sections on the Great Bank of Guizhou, an isolated Late Permian to Middle Triassic carbonate platform in the Nanpanjiang Basin of southern China. This volatile Early Triassic carbon isotope record stabilized in the Middle Triassic. Abbreviations: P-TB, Permian–Triassic boundary; Changhs., Changhsingian; Griesbach., Griesbachian; Smit., Smithian; Spath., Spathian; L. Perm., Late Permian; Trias., Triassic. From Payne and Kump (2007). Reproduced with permission from Elsevier.

between the Siberian Trap eruptions and conditions leading to deposition of these sediment types. These stressful environmental conditions also led to the restriction of metazoan reef building, including the lack of corals, during the Early Triassic (Fig. 13.7). Subsequent studies have shown that metazoan reef building was less constrained during this time, with bivalves appearing in microbial buildups at the end of the Early Triassic (Fig. 13.8) and sponges reappearing as components of buildups with microbialites in the Smithian (Fig. 13.9). Ocean acidification, as indicated by trends of calcium isotopes, may have contributed to the demise of metazoan reefs during the end-Permian mass extinction (Fig. 13.10).

The unusually warm oceans during this time were characterized by a widespread oxygen minimum zone (OMZ), which, beginning with the mass extinction event, periodically impinged from deep into shallow environments (Fig. 13.11). These shallow anoxic waters were rich in hydrogen sulfide, as indicated by evidence of green sulfur bacteria, indicating euxinia in the photic zone. Along with increased temperatures, anoxia and sulfide toxicity are hypothesized as contributing factors to the end-Permian mass extinction and prolonged recovery. Stressful environmental conditions, including low oxygen conditions, may have contributed to a reduction in depth (Fig. 1.3) and extent (Fig. 13.12) of bioturbation by infaunal organisms during much of the Early Triassic. Predominance of low oxygen waters has been postulated from the start of the Late Permian until the Middle Triassic (Fig. 13.13). These extensive low oxygen ocean environments would have been a major contributor to the stresses that caused the end-Permian mass extinction and which changed the overall nature of the marine biota from one that was largely nonmobile (Paleozoic Fauna) to one that was self-mobile (Modern Fauna) (Fig. 13.13). An onshore–offshore study of steno-laemate bryozoan occurrences from the Middle Permian to the Jurassic shows an extinction pattern suggesting that an offshore stress, such as low oxygen water, impinged on the shelf beginning in the Middle Permian, ultimately resulting in encroachment onto the shelf at the end of the Permian and the resulting mass extinction (Fig. 13.14).

The effects of the mass extinction on land greatly increased weathering and decreased plant life, leading to a significant increase in supply of terrigenous sediments to shallow-marine systems (Fig. 13.11). Pulses of increased temperature led to increased biotic crises in the Early Triassic (Fig. 13.15). Similarly, other shallow seafloor environments were characterized by low biodiversity and cosmopolitan taxa, particularly in the beginning of the Early Triassic (Induan) (Figs. 13.16–13.18). A number of Early Triassic taxa have been reported to exhibit a reduction in size from the Permian into the Triassic (Figs. 13.19 and 13.20). Whether this is a global phenomenon or the product of regional environmental conditions is still a subject of ongoing research.

In particular, the significance of increased temperature, low oxygen ocean water development,

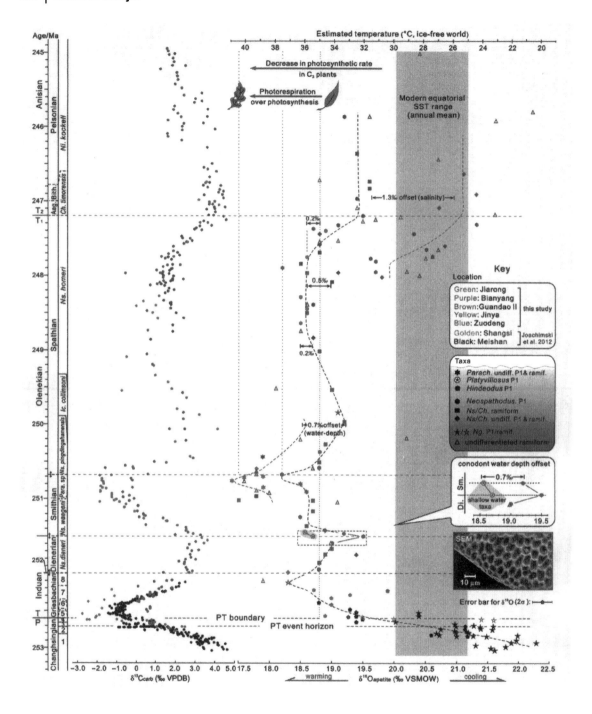

Figure 13.3 Sea surface temperature trends from the Late Permian into the Middle Triassic. These have been determined from oxygen isotopes of conodont apatite from the Nanpanjiang Basin of southern China and are compared with carbon isotopes of associated carbonates (Fig. 13.2). The estimated sea surface temperatures (SST) for this Permian–Triassic equatorial area are also compared with modern SST. Note near-synchronous perturbations in the records for both carbon and oxygen isotopes. Oxygen isotopes indicate a significant warming through the Permian–Triassic boundary transition, two thermal maxima in the late Griesbachian and late Smithian, cooling in the early Spathian followed by temperature stabilization, and further cooling and stabilization into the Middle Triassic. In the key, location of data points of different colors is indicated, and these can be identified in the color insert; differently shaped icons show the different conodont taxa which were sampled for data points; and conodonts which lived at different water depths show offsets in estimated temperatures. Scanning electron microscope (SEM) analysis of conodont surfaces shows microreticulation and no sign of recrystallization, indicating a primary signal was obtained, as exemplified by the conodont surface SEM. At temperatures greater than 35 °C, photorespiration predominates over photosynthesis, and above 40 °C, most plants perish; leaf icons represent marine and terrestrial C3 plants. The occurrence of the end-Permian mass extinction (PT Event Horizon) occurs earlier than the biostratigraphically defined Permian–Triassic boundary (PT boundary) (Fig. 13.9). Aeg., Aegean; Bit., Bithynian; for conodont zonations, see original reference for abbreviations. Standards for isotope measurements: VSMOW, Vienna Standard Mean Ocean Water; VPDB, Vienna Pee Dee Belemnite. From Sun et al. (2012). Reproduced with permission from the American Association for the Advancement of Science. (*See insert for color representation.*)

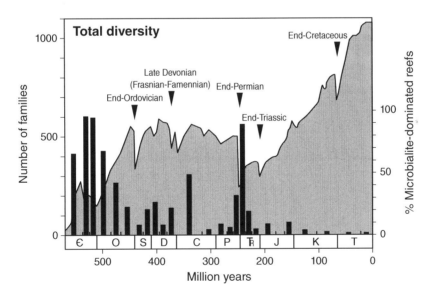

Figure 13.4 Abundance of microbialites in the Early Triassic. This shows Phanerozoic marine familial diversity as determined by Sepkoski (Fig. 3.11) overlain by the relative abundance of microbialite-dominated reefs through the Phanerozoic, binned by epoch. The locations of the traditional "Big Five" mass extinctions are indicated. Note the drop-off from high levels of microbialite-dominated reefs in the early Paleozoic to relatively low levels after the Ordovician except for prominent peaks of abundance in the Carboniferous following the Late Devonian mass extinction and in the Early Triassic following the end-Permian mass extinction. Microbialite-dominated reef abundance data includes reef mounds, mud mounds, biostromes, and framework-dominated rigid reefs in which microbialite is volumetrically the most prevalent biotic component. From Mata and Bottjer (2011). Reproduced with permission from John Wiley & Sons.

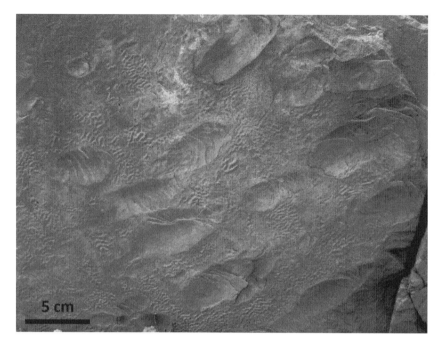

Figure 13.5 Lower Triassic (Spathian) wrinkle structures associated with unidirectional scour marks (scoop-shaped impressions) on a bedding plane of the Virgin Limestone Member (Moenkopi Formation). This bedding plane, found near Overton in the Muddy Mountains, Nevada (United States), forms the top of a calcareous siltstone. Wrinkle structures have crest heights of 1–2 mm, and associated sedimentary features include ripple marks and cross-bedding. Additional information can be found in Pruss et al. (2004). Photograph by Scott A. Mata. Reproduced with permission. (*See insert for color representation.*)

and ocean acidification for affecting different components of the marine fauna can be further understood. Increased ocean temperatures led to a restriction of many components of the biota to cooler temperatures of high latitudes (Fig. 13.21). Assessment of ecosystem recovery shows that at different locations and times ecosystems were recovering at different rates (Fig. 13.22). Bryozoans are also restricted to high northern latitudes during the Early Triassic, which may similarly have been a reaction to increased ocean temperatures (Fig. 13.23). An oxygenated nearshore "habitable zone" has been proposed as an environmental refuge for benthic marine organisms during times when oxygen-deficient waters predominated in Early Triassic oceans (Fig. 13.24).

As an understanding of this variable temporal and geographic mosaic of benthic ecological recovery has developed, methods to assess the extent of ecological recovery in benthic environments during the Early Triassic have become increasingly necessary (e.g., Fig. 13.25), utilizing evidence from body fossils, trace fossils, microbialites, and reef structures. This sort of approach will eventually lead to a detailed mapping in space and time of benthic ecological structure and recovery in the Early Triassic. As outlined in this section, our current understanding is that benthic organisms show a variable but overall slow recovery through the Early Triassic.. In contrast, pelagic organisms such as ammonoids and conodonts appear to have had pulses of rapid recovery followed by intervals

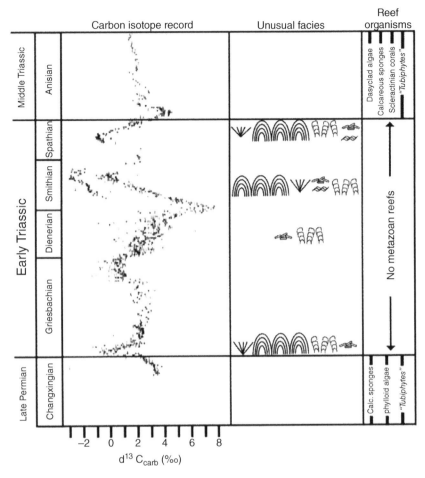

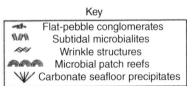

Figure 13.6 Distribution of microbialites and other Early Triassic unusual sedimentary facies, termed anachronistic facies, compared with the carbon isotope record and reef organism trends. Microbialites include stromatolites in subtidal microbialites and microbial patch reefs, as well as wrinkle structures. Flat-pebble conglomerates form when carbonate surface sediments are lithified and then broken up into conglomerates and redeposited, due to storm processes; reduced levels of bioturbation are interpreted to enhance the lithification process. Carbonate seafloor precipitates form when seawater supersaturated in calcium carbonate leads to the growth of aragonite fans on or just below the seafloor. Carbonate isotope data as shown in Fig. 13.2. This compilation was done before several types of metazoans were found in Lower Triassic microbialite reefs (Figs. 13.8 and 13.9). From Pruss et al. (2006). Reproduced with permission from Elsevier.

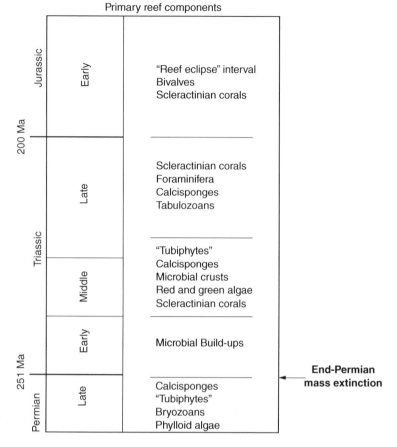

Figure 13.7 Dominant reef components from the Late Permian through the Early Jurassic. "Tubiphytes" is a problematic fossil that has been interpreted as a variety of organisms, including sponges. A "reef eclipse" interval is a time when reefs occur relatively rarely. From Pruss and Bottjer (2005). Reproduced with permission from Elsevier.

of decline during this time interval (Fig. 13.26). Continued investigation of Early Triassic benthic and pelagic ecosystems will illuminate the reasons for these differences.

The lethal effects of increased temperature can operate over millions of years, while the lethal effects of ocean acidification and increased anoxia/euxinia in shallow waters typically operate over shorter time intervals. Thus, it is likely that the sustained lack of corals during the 5-million-year-long Early Triassic was not primarily due to acidification or increased anoxia/euxinia in ocean ecosystems, but increased ocean temperatures. Increased ocean temperatures may have also acted as the mechanism for the reduced size of organisms in Early Triassic ocean ecosystems.

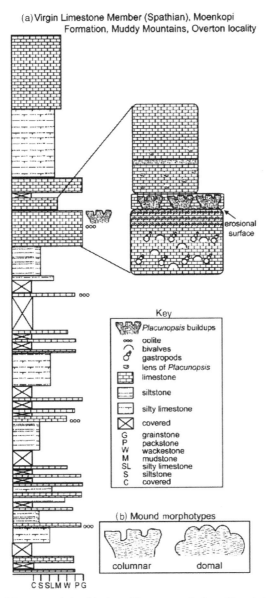

Figure 13.8 Lower Triassic buildups containing bivalves. These mounds from Nevada include the bivalve *Placunopsis*. (A) Measured stratigraphic section of the Spathian (latest Early Triassic) Virgin Limestone member, Moenkopi Formation, Muddy Mountains Overton locality, showing location of bivalve buildups in the upper half of the section. (B) Mound morphotypes in outcrop. From Pruss et al. (2007). Reproduced with permission from the SEPM Society for Sedimentary Geology.

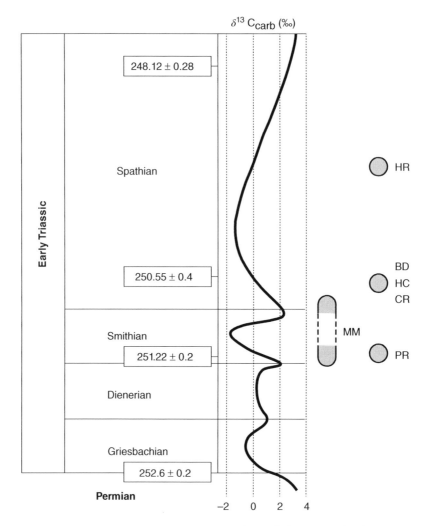

Figure 13.9 Stratigraphic occurrence in Nevada and Utah of Lower Triassic bioaccumulations and reefs which include sponges, serpulids, microbialites, and other benthic eukaryotic organisms. Age relationships are indicated by absolute ages and a simplified global carbon isotope trend. Locations of study areas, marked by gray circles, include HR, Humboldt Range; BD, Beaver Dam Mountains; HC, Hurricane Cliffs; CR, Confusion Range; MM, Mineral Mountains; PR, Pahvant Range. From Brayard et al. (2011). Reproduced with permission from Macmillan Publishers Ltd.

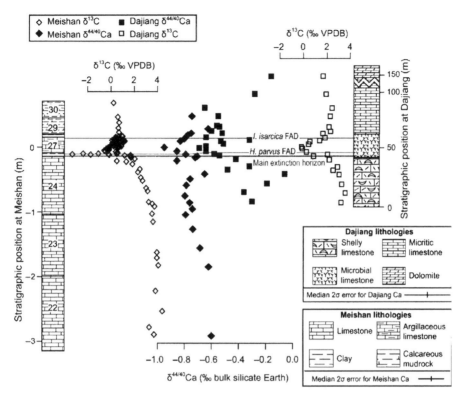

Figure 13.10 Evidence for ocean acidification during the end-Permian mass extinction. This is indicated by calcium and carbon isotope data from Meishan and Dajiang, China. The main extinction horizon at Meishan for the end-Permian mass extinction is placed in Bed 25, while the Permian–Triassic boundary is defined by the first appearance of the conodont *Hindeodus parvus*, which at Meishan falls in Bed 27c. These points, along with the first appearance datum (FAD) of *Isarcicella isarcica*, were used to correlate the Meishan section with previously reported calcium isotope data from Dajiang. Both sections exhibit a negative calcium isotope excursion above the main extinction horizon, which is best accounted for as due to ocean acidification. Both sections exhibit positive correlations between the calcium isotope and carbon isotope records. See also Clarkson et al. (2015) for further evidence, from a study of boron isotopes, of ocean acidification during this time. From Hinojosa et al. (2012). Reproduced with permission from the Geological Society of America.

(a) Late Permian

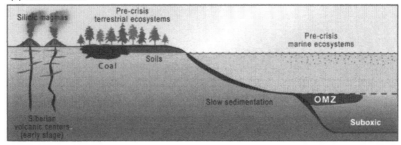

(b) End-Permian event

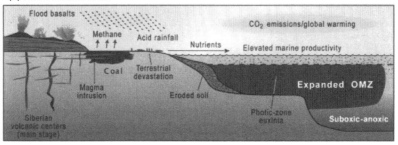

(c) Early Triassic

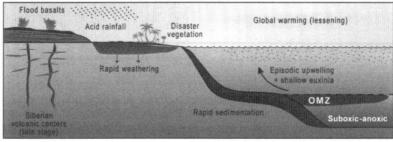

(d) Middle Triassic

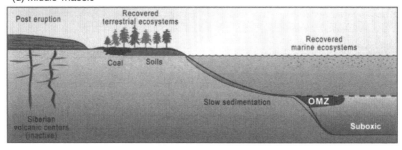

Figure 13.11 Interpretive reconstructions of terrestrial–marine teleconnections from the Late Permian into the Middle Triassic. (a) Early-stage Siberian Traps volcanism with minimal environmental effects during the Late Permian. (b) Main stage eruptions with attendant environmental effects during the latest Permian. (c) Late-stage eruptions with lessening environmental effects during Early Triassic. (d) Posteruption recovery of terrestrial and marine ecosystems in the Middle Triassic. From Algeo et al. (2011). Reproduced with permission from Elsevier. (*See insert for color representation.*)

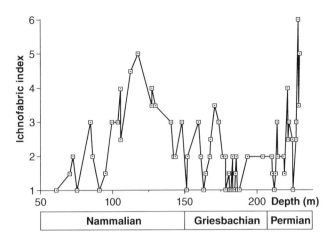

Figure 13.12 Measurements of ichnofabric in the Gartnerkofel-1 core using the ichnofabric index method (Fig. 5.4). This shows a drop in measured ichnofabric index from the Permian into the Griesbachian, with subsequent fluctuations in the Nammalian. The Nammalian is equivalent to the Dienerian and Smithian stages of the Early Triassic. From Twitchett and Wignall (1996). Reproduced with permission from Elsevier.

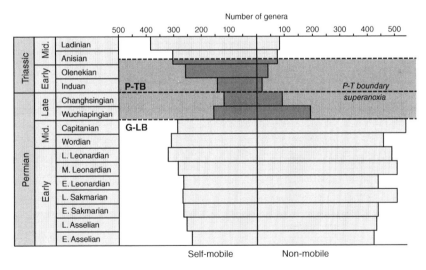

Figure 13.13 Low oxygen conditions and change in biodiversity of self-mobile and nonmobile genera across the end-Permian mass extinction. Analyses of stratigraphic sections in Japan and elsewhere deposited in deep-sea environments indicate long-term prevailing low oxygen conditions (superanoxia) from the Middle/Late Permian (Guadalupian–Lopingian; G-LB) boundary to the beginning of the Middle Triassic. Extinction first began at the G-LB boundary, culminating at the Permian–Triassic boundary (P-TB), and environmental stress continued until the Middle Triassic. Preferential extinction of nonmobile sessile benthic organisms occurred during this interval, with preferential survival of self-mobile organisms. From Isozaki (2007). Reproduced with permission from Springer Science and Business Media.

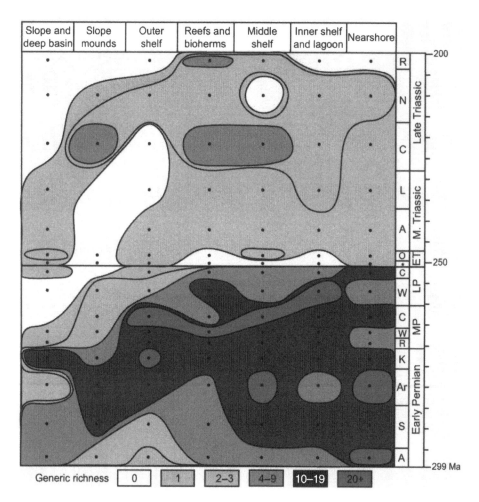

Figure 13.14 Contoured Permian–Triassic time–environment (T–E) diagram of marine stenolaemate bryozoans. Each dot represents a data point, either for an assemblage with the greatest bryozoan generic richness in each T–E bin or for absence of bryozoans validated by the taphonomic control group rhynchonelliform brachiopods). Note that bryozoan generic richness began to decrease in offshore environments in the Middle Permian, and this pattern of generic decrease continued into the Late Permian in more onshore environments. The end-Permian mass extinction then marked a geologically sudden extinction across all environments. Bryozoan generic richness was then very low until the Late Triassic when it began to increase in slope mound, reef, bioherm, and middle shelf environments. From Bottjer et al. (2008). Reproduced with permission from the Geological Society of America. (*See insert for color representation.*)

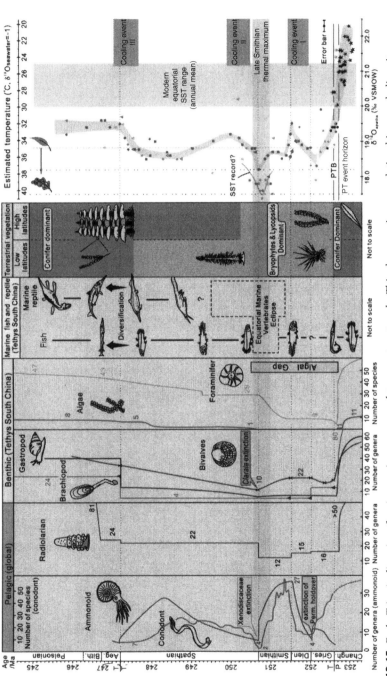

Figure 13.15 Early Triassic diversity of major marine groups and temperature trends. This shows an inverse relationship – peak diversity corresponds to cool climate conditions around the Dienerian–Smithian boundary, early Spathian, and early Anisian (named cooling events I–III), whereas low diversity in Griesbachian and Smithian correlates with peak temperatures. Fish and marine reptiles only show the general presence of taxa; no quantitative diversity data are available. Floral data show the loss of equatorial conifer-dominated forests above the Permian–Triassic (PT) boundary, with the earlier reappearance of this forest type at high latitudes. For estimated temperature (Fig. 13.3), vertically trending gray band represents the first-order seawater temperature trend (upper water column, ~70 m water depth), whereas the solid line indicated by "SST record?" represents possible sea surface temperatures derived from shallow water conodont taxa. Same stratigraphic scheme as in Fig. 13.3. From Sun et al. (2012). Reproduced with permission from the American Association for the Advancement of Science. (*See insert for color representation.*)

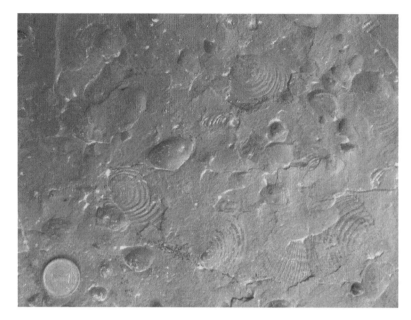

Figure 13.16 A characteristic Early Triassic (Induan) benthic assemblage consisting of the bivalves *Claraia* and *Unionites* on a bedding plane from the Lower Siusi Member of the Werfen Formation, northern Italy. *Claraia*, 2–4 cm in greatest dimension, has concentric growth lines, and *Unionites* has a smoother more triangular shell. For life habits of *Claraia* and *Unionites*, see Fig. 13.17. From the type location of the Siusi Member due south of the town of Siusi in the Dolomites of Alto Adige (South Tyrol). Photograph by Roger Twitchett. Reproduced with permission. (*See insert for color representation.*)

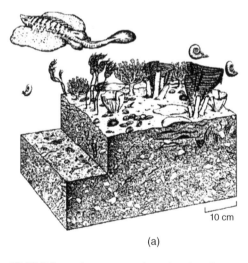

(a)

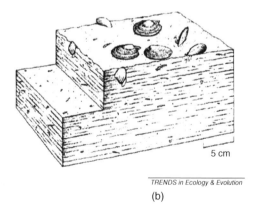

5 cm

TRENDS in Ecology & Evolution

(b)

Figure 13.17 Schematic reconstructions showing change in characteristic benthic paleocommunities from the Late Permian to the Early Triassic in southern China. These reconstructions show the ancient seabed in southern China immediately before (a) and after (b) the end-Permian mass extinction. Note the richness of benthic life before the crisis and the absence of such species after. A Permian marine fauna of 100 or more species was reduced to an assemblage consisting of a few Early Triassic taxa, represented here by the epifaunal *Claraia* and infaunal *Unionites*, much like that shown for Italy in Fig. 13.16, as well as the byssally attached *Promyalina*. From Benton and Twitchett (2003). Reproduced with permission from Elsevier.

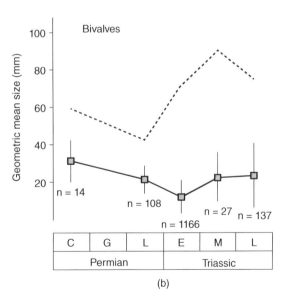

Figure 13.18 Early Triassic disaster taxa. These include (a) the brachiopod *Lingularia* and the bivalves (b) *Claraia*, (c) *Eumorphotis*, (d) *Unionites*, and (e) *Promyalina*. These taxa are typically the most abundant in benthic communities worldwide, particularly in the earliest part of the Early Triassic. From Benton (2003). Reproduced with permission from Tony Hallam and Paul Wignall.

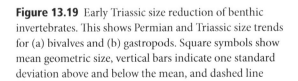

Figure 13.19 Early Triassic size reduction of benthic invertebrates. This shows Permian and Triassic size trends for (a) bivalves and (b) gastropods. Square symbols show mean geometric size, vertical bars indicate one standard deviation above and below the mean, and dashed line shows maximum geometric size. Abbreviations: C, Cisuralian; G, Guadalupian; L, Lopingian; E, Early; M, Middle; L, Late. Twitchett (2007). Reproduced with permission of Elsevier.

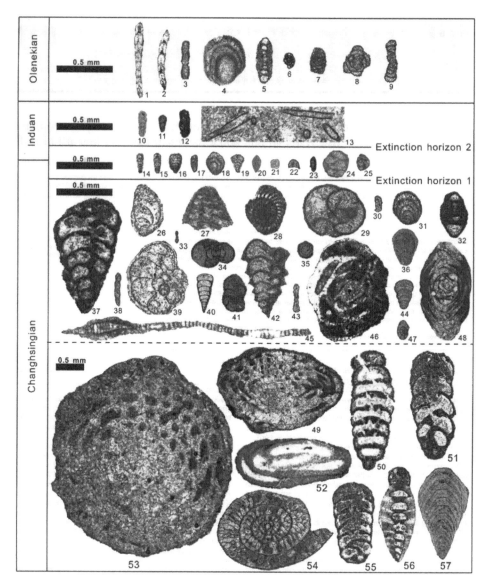

Figure 13.20 Early Triassic size reduction of foraminifera. This shows selected foraminifera showing test size distributions for Permian (Changhsingian) into Early Triassic (Induan and Olenekian). (1) *Nodosaria*; (2) *Dentalina*; (3) indeterminate Lagenida genus; (4) *Ichthyofrondina*; (5) *Arenovidalina*; (6) *Meandrospira*; (7) *Pilammina*; (8) *Glomispira*; (9) *Glomospirella*; (10) *Lingulina*; (11) *Nodosinelloides*; (12) *Glomospira*; (13) *Earlandia*; (14) *Nodosinelloides*; (15) *Nodosaria*; (16) *Geinitzina*; (17) *Frondina*; (18) *Ichthyofrondina*; (19) *Cryptoseptida*; (20) *Robuloides*; (21) *Rectostipulina*; (22) *Diplosphaerina*; (23) *Neoendothyra*; (24) *Globivalvulina*; (25) *Hemigordius*; (26) *Calvezina*; (27) *Tetrataxis*; (28) *Reichelina*; (29) *Paraglobivalvulina*; (30) *Ammovertella*; (31) *Ichthyofrondina*; (32) *Neoendothyra*; (33) *Pseudoammodiscus*; (34) *Postendothyra*; (35) *Biorbis*; (36) *Cryptomorphina*; (37) *Paleotextularia*; (38) *Vervilleina*; (39) *Partisania*; (40) *Protonodosaria*; (41) *Sengoerina*; (42) *Dagmarita*; (43) ?*Tournayella*; (44) "*Pseudoglandulina*"; (45) *Parareichelina*; (46) *Glomomidiella*; (47) *Pseudonovella*; (48) *Multidiscus*; (49) *Nankinella*; (50) *Cribrogenerina*; (51) *Deckerella*; (52) *Agathammina*; (53) *Pisolina*; (54) *Paleofusilina*; (55) *Climacammina*; (56) *Pachyphloia*; and (57) *Colaniella*. Song et al. (2011). Reproduced with permission of Elsevier.

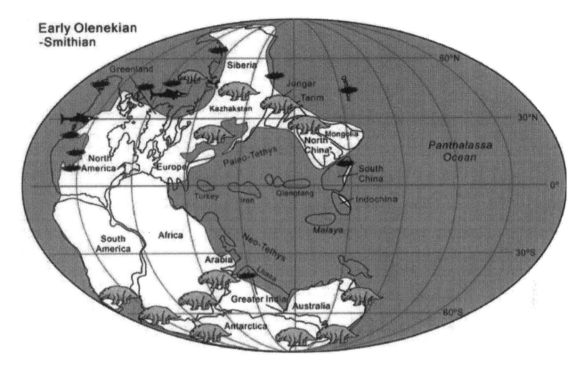

Figure 13.21 Smithian restriction of tetrapods, marine reptiles, and fish to polar regions. The Smithian was the hottest time in the Early Triassic and possibly in the entire Phanerozoic. Tetrapods indicated by gray symbols, fish and marine reptiles indicated by black symbols. Sun et al. (2012). Reproduced with permission of the American Association for the Advancement of Science. (*See insert for color representation.*)

Figure 13.22 Different Early Triassic regional patterns of temporal recovery from the end-Permian mass extinction. These are shown from the (a) western United States; (b) northern Italy; and (c) Oman. Numbers 1–4 indicate recovery stages as determined by biodiversity and paleoecological data. Arrow indicates onset of shallow-marine anoxia in Oman in the basal Dienerian. Twitchett et al. (2004). Reproduced with permission of the Geological Society of America.

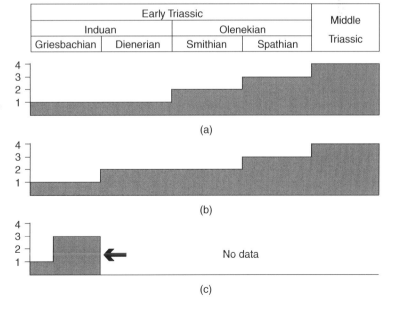

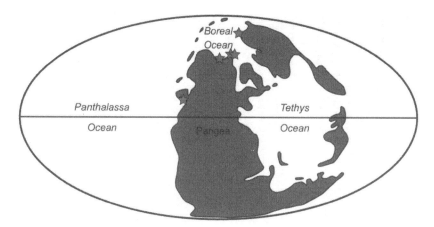

Figure 13.23 Paleobiogeographic distribution of Early Triassic bryozoans. These are indicated by stars, note concentration at north polar region. Bottjer et al. (2008). Reproduced with permission of the Geological Society of America. (*See insert for color representation.*)

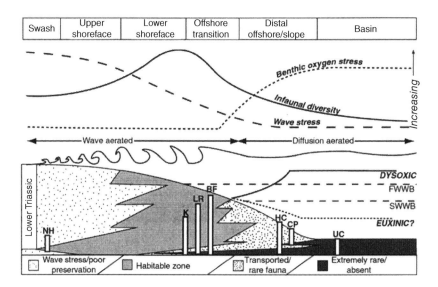

Figure 13.24 The Early Triassic habitable Zone. This is depicted with respect to shoreface position, infaunal diversity peak, and oxygen and wave stress (top). Schematic cross section of typical shoreline and relative environmental position of studied Canadian stratigraphic sections from northwest Pangea (bottom); Abbreviations: NH, North Hamilton Peninsula; K, Kahntah Field; LR, Liard River; BF, Borup Fjord; HC, Hood Creek; CP, Confederation Point; UC, Ursula Creek. FWWB is fair-weather wave base; SWWB is storm-weather wave base. Beatty et al. (2008). Geology. Reproduced with permission of the Geological Society of America.

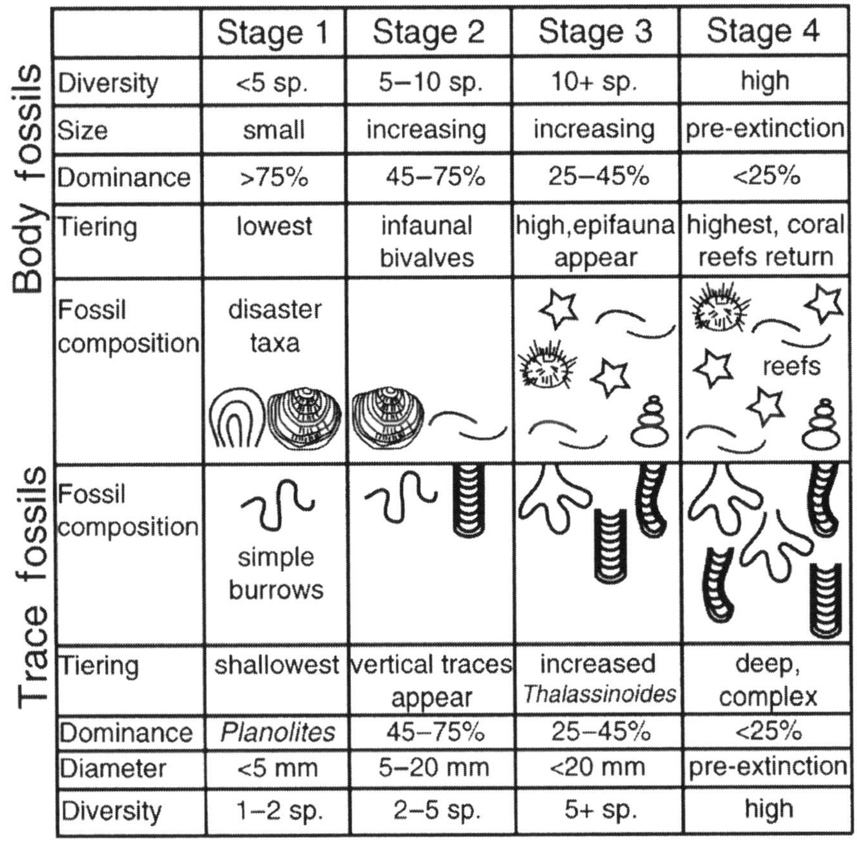

		Stage 1	Stage 2	Stage 3	Stage 4
Body fossils	Diversity	<5 sp.	5–10 sp.	10+ sp.	high
	Size	small	increasing	increasing	pre-extinction
	Dominance	>75%	45–75%	25–45%	<25%
	Tiering	lowest	infaunal bivalves	high, epifauna appear	highest, coral reefs return
	Fossil composition	disaster taxa			reefs
Trace fossils	Fossil composition	simple burrows			
	Tiering	shallowest	vertical traces appear	increased *Thalassinoides*	deep, complex
	Dominance	*Planolites*	45–75%	25–45%	<25%
	Diameter	<5 mm	5–20 mm	<20 mm	pre-extinction
	Diversity	1–2 sp.	2–5 sp.	5+ sp.	high

Figure 13.25 Ecological recovery model for the Early Triassic used to grade the level of recovery after the end-Permian mass extinction. This schematic represents the four stages that comprise this model, from low level of recovery (Stage 1) to complete ecological recovery (Stage 4). The four stages are determined through analysis of body fossils and microbialites including diversity, size, dominance ratios, tiering height and depth, fossil composition, and presence of coral reefs, as well as trace fossil and bioturbation diversity, burrow size, dominance, burrow depth, tiering relationships, and trace fossil composition. For a fossil deposit to be considered for a given recovery stage, it must meet 50% or more of the diversity, size, dominance, or burrow depth characteristics and must contain evidence for epifaunal and/or infaunal tiering depending on available data. For body fossils, fossil composition symbols include stromatolites and *Claraia* as Stage 1 disaster taxa, Stage 2 *Claraia* and bivalves, Stages 3 and 4 bivalves, gastropods, echinoids, and crinoids (stars). For trace fossils, fossil composition includes *Planolites* as Stage 1 simple burrows, Stage 2 *Planolites* and *Diplocraterion*, Stage 3 *Diplocraterion*, *Rhizocorallium*, and *Thalassinoides* (branched), and Stage 4 deep and complex burrows. This rubric is modified from Twitchett et al. (2004), used in the analysis shown in Fig. 13.22, and Twitchett (2006). Body fossil dominance is based on Clapham et al. (2006) and Clapham and Bottjer (2007). Trace fossil diameter is based on Zonneveld et al. (2010). Pietsch and Bottjer (2014). Reproduced with permission of Elsevier.

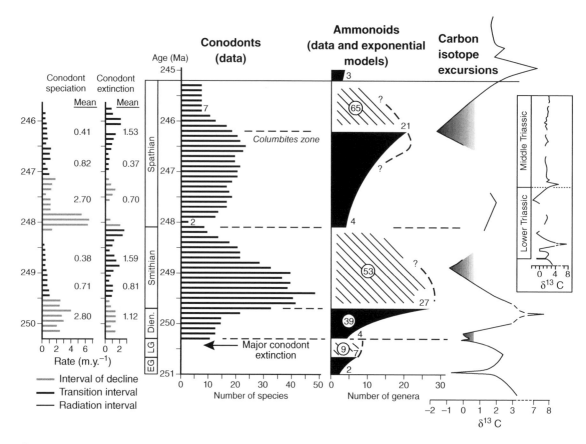

Figure 13.26 Simultaneous radiations and mass extinctions of conodonts and ammonoids as well as negative carbon isotope excursions during intervals of Early Triassic mass extinction. Numbers of conodont species are three-interval moving averages for 100,000-year intervals based on a global compilation from which rates of speciation and extinction were also calculated; uncertain range extensions were not included. Bold numbers give standing global diversities of ammonoid genera at the end of the Griesbachian and Dienerian and the top of the *Columbites* zone. Encircled numbers give total diversity for the post-*Columbites* zone interval and for earlier intervals. For hatched intervals, the exact pattern of change for ammonoid diversity is uncertain; for solid gray intervals, diversification is modeled as exponential but for the Spathian, the rate may have begun to decline earlier than shown (see dashed line), paralleling the pattern for conodonts. In bold are numbers of conodont species and ammonoid genera at ends of mass extinction intervals. The Early Triassic carbon isotope graph is simplified from a composite plot for data from four stratigraphic sections in southern China, the curve for the Middle Triassic (inset) represents a single section in southern China. From Stanley (2009). Reproduced with permission from the National Academy of Sciences.

End-Cretaceous mass extinction

The mass extinction at the end of the Cretaceous provides a different set of conditions from which to examine the ecological response to a mass extinction event. The extraterrestrial impact at the end of the Cretaceous led to extensive extinctions on land and in the sea. As reflected in the boundary clay layer, this impact threw an enormous amount of debris into the atmosphere, which is hypothesized to have blocked out sunlight for a year or more. A prediction of this scenario is that there would be a dramatic decline in primary productivity and hence abundance of phytoplankton in marine ecosystems after the impact. It would also be predicted that for encrusting bryozoans, the cyclostomes, which require less nutrients than cheilostomes, would increase dramatically in abundance after the impact. But this is not what is seen in a paleoecological study of cyclostome and cheilostome encrusting bryozoans from boundary localities around the world, such as Stevns Klint (Fig. 13.27), which shows only a small cyclostome increase. Similarly, within cyclostomes, the colony form composed of uniserial branching runners is thought to be favored in low nutrient environments, but they show no change across the boundary (Fig. 13.27). Thus, this analysis of encrusting bryozoans shows that in marine ecosystems strong supporting evidence for a collapse in productivity is not borne out.

As shown for the end-Permian mass extinction, variability in space and time for the ecological recovery from the end-Cretaceous mass extinction can be studied as well. This recovery variability can be seen through study of microfossils from open ocean sites where the record includes the latest Cretaceous and the beginning of the Cenozoic. Plots of calcareous sediment flux, flux of foraminifera, and percent of foraminiferal-sized grains, which reflect aspects of pelagic community structure, show an alternative community structure existed in pelagic ecosystems for several hundred thousand years after the mass extinction. These plots demonstrate that within the recovery interval there was geographic heterogeneity that also played out temporally from different sites ranging from the North Pacific to the South Atlantic (Fig. 13.28).

Comparative ecological change of mass extinctions

Mass extinctions can be ranked by the amount of biodiversity loss that occurs. They also can be viewed by how much ecological change occurs. Ranking the scale of ecological change has been attempted through characterization of a series of paleoecological levels. As defined in Table 13.1, the first level encompasses changes that occur between ecosystems, the second level is changes that occur within ecosystems, the third level includes changes that occur within community types, and the fourth level includes changes within communities.

Using paleoecological levels, one can rank the ecological severity of each mass extinction (Table 13.2). A surprising result of this analysis is that the ranking of mass extinctions by taxonomic biodiversity loss does not match a ranking done for analysis of ecosystem degradation (Table 13.2). This demonstrates that the amount of biodiversity loss is not indicative on a one-to-one basis of the amount of ecological degradation that occurs during a mass extinction. It also demonstrates that different taxa in an ecosystem have different ecological values and that the ecological values of the organisms that become extinct can vary significantly.

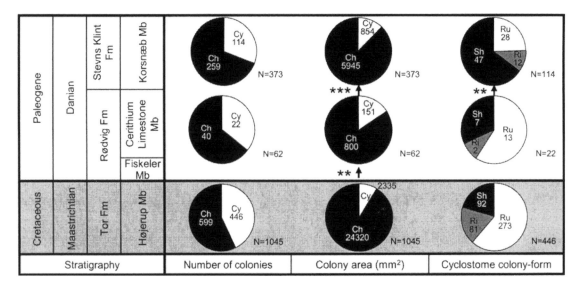

Figure 13.27 Change in encrusting bryozoans across the Cretaceous–Paleogene boundary. This diagram shows for each stratigraphic interval across the boundary at Stevns Klint, Denmark, proportions in numbers of collected encrusting cyclostome and cheilostome colonies, changes in total area of all studied cyclostome and cheilostome colonies, and changes in colony form. There is no statistical difference in numbers of cheilostome and cyclostome colonies in each of the studied stratigraphic intervals. There is a small but statistically significant increase in cyclostome colony area across the boundary and a statistically significant small decrease in cyclostome colony area from the Cerithium Limestone Member to the Korsnaeb Member. Cheilostomes only occur as encrusting sheet colonies in this study, so only cyclostome colony form variety was studied. Here, there is no change in cyclostome colony form across the boundary. A large and statistically significant change in colony form occurs from the Cerithium Limestone Member to the Korsnaeb Member from dominance by runners to dominance by sheets. Similar patterns are shown in studies of encrusting bryozoans across the Cretaceous–Paleogene boundary in Mississippi and Alabama. Absolute numbers of colonies and total clade area are shown. Abbreviations: Ch, cheilostome; Cy, cyclostome; Ru, runner; Ri, ribbon; Sh, sheet; Fm, formation; Mb, member. Arrows indicate significant changes ($**p < 0.01$; $***p < 0.001$). From Sogot et al. (2013). Reproduced with permission from the Geological Society of America.

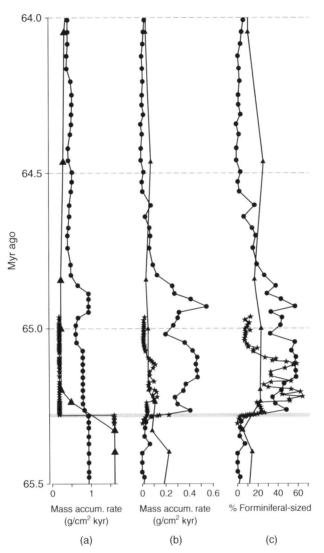

Figure 13.28 Regional variability in recovery of pelagic communities from the end-Cretaceous mass extinction (∼65.3 million years ago). Samples from deep-sea cores were studied from Shatsky Rise, North Pacific (ODP Site 1209, solid circles); Walvis Ridge, eastern South Atlantic (ODP Site 1262, asterisks); and Sao Paolo Plateau, western South Atlantic (DSDP Site 356, triangles). From these samples, the composition of the sediment was studied to calculate a, flux of calcareous sediment; b, foraminiferal flux; and c, % foraminiferal-sized grains. The difference between the calcareous flux and the foraminiferal flux is the flux of calcareous nannofossils, primarily coccolithophorids. From Hull et al. (2011). Reproduced with permission from Macmillan Publishers Ltd.

Table 13.1 Definition of paleoecological levels and characteristic signals for each level (from Droser et al. 2000). Bambachian megaguilds are defined in Fig. 12.11.

Level	Definition	Signals
First	Appearance/ disappearance of an ecosystem	1. Initial colonization of environment
Second	Structural changes within an ecosystem	1. First appearance of, or changes in ecological dominants of higher taxa 2. Loss/appearance of metazoan reefs 3. Appearance/disappearance of Bambachian megaguilds
Third	Community-type level changes structure	1. Appearance and/or disappearance of within an established ecological community types 2. Increase and/or decrease in tiering complexity 3. "Filling-in" or "thinning" within Bambachian megaguilds
Fourth	Community-level changes	1. Appearance and/or disappearance of communities 2. Taxonomic changes within a clade

Source: From McGhee et al. (2013). Reproduced with permission from Elsevier.

Table 13.2 Ecological-severity rankings of the traditional 11 largest Phanerozoic biodiversity crises since the beginning of the Ordovician compared with taxonomic-severity rankings. Ecological-severity rankings were established using the system of paleoecological levels, and each biodiversity crisis is named by the stage marking the crisis; the traditional names for the "Big Five" biodiversity crises (Fig. 13.4) are indicated in parentheses.

Ecological-severity ranking	Taxonomic-severity ranking
1. Changhsingian (end-Permian)	1. Changhsingian
2. Maastrichtian (end-Cretaceous)	2. Rhaetian
3. Rhaetian (end-Triassic)	3. Hirnantian
4. Frasnian (Late Devonian)	4. Famennian
5. Capitanian	5. Maastrichtian, Frasnian
6. Serpukhovian	6. Serpukhovian
7. Famennian, Hirnantian (end-Ordovician)	7. Givetian
8. Givetian	8. Eifelian
9. Eifelian, Ludfordian	9. Capitanian
	10. Ludfordian

Source: From McGhee et al. (2013). Reproduced with permission from Elsevier.

Summary

Periods of abrupt geological change affecting biodiversity have changed the course of evolution through mass extinction. Paleoecological studies, a number of which have been presented in this chapter, show how these perturbations have also strongly affected the trajectory of ecosystems. It is well known that Earth is currently in a period of biodiversity loss that may result in another great mass extinction. For this reason, studies of ancient mass extinctions have been particularly compelling in order to learn more about the possible consequences of the current situation. This area of research will be explored further in the following chapter.

References

Algeo, T.J., Chen, Z.Q., Fraiser, M.L. & Twitchett, R.J. 2011. Terrestrial-marine teleconnections in the collapse and rebuilding of Early Triassic marine ecosystems. *Palaeogeography, Palaeoclimatology, Palaeoecology* 308, 1–11.

Beatty, T.W., Zonneveld, J.-P. & Henderson, C.M. 2008. Anomalously diverse Early Triassic ichnofossil assemblages in northwest Pangea: A case for a shallow-marine habitable zone. *Geology* 36, 771–774.

Benton, M.J. 2003. *When Life Nearly Died: The Greatest Mass Extinction of All Time*. Thames & Hudson.

Benton, M.J. & Twitchett, R.J. 2003. How to kill (almost) all life: the end-Permian extinction event. *TRENDS in Ecology and Evolution* 18, 358–365.

Bottjer, D.J., Clapham, M.E., Fraiser, M.L. & Powers, C.M. 2008. Understanding mechanisms for the end-Permian mass extinction and the protracted Early Triassic aftermath and recovery. *GSA Today* 18, 4–10.

Brayard, A., Vennin, E., Olivier, N., Bylund, K.G., Jenks, J., Stephen, D.A., Bucher, H., Hofmann, R., Goudemand, N. & Escarguel, G. 2011. Transient metazoan reefs in the aftermath of the end-Permian mass extinction. *Nature Geoscience* 4, 693–697.

Clapham, M.E., Bottjer, D.J., Powers, C.M., Bonuso, N., Fraiser, M.L., Marenco, P.J., Dornbos, S.Q. & Pruss, S.B. 2006. Assessing the ecological dominance of Phanerozoic marine invertebrates. *Palaios* 21, 431–441.

Clapham, M.E. & Bottjer, D.J. 2007. Prolonged Permian–Triassic ecological crisis recorded by molluscan dominance in Late Permian offshore assemblages. *Proceedings of the National Academy of Sciences* 104, 12971–12975.

Clarkson, M.O., Kasemann, S.A., Wood, R.A., Lenton, T.M., Daines, S.J., Richoz, S., Ohnemueller, F., Meixner, A., Poulton, S.W. & Tipper, E.T. 2015. Ocean acidification and the Permo-Triassic mass extinction. *Science* 348, 229–232.

Droser, M.L., Bottjer, D.J., Sheehan, P.M. & McGhee, G.R., Jr. 2000. Decoupling of taxonomic and ecologic severity of Phanerozoic mass extinctions. *Geology* 28, 675–678.

Hinojosa, J.L, Brown, S.T., Chen, J., DePaolo, D.J., Paytan, A., Shen, S.-Z. & Payne, J.L. 2012. Evidence for end-Permian ocean acidification from calcium isotopes in biogenic apatite. *Geology* 40, 743–746.

Hull, P.M., Norris, R.D., Bralower, T.J. & Schueth, J.D. 2011. A role for chance in marine recovery from the end-Cretaceous extinction. *Nature Geoscience* 4, 856–860.

Isozaki, Y. 2007. Plume Winter Scenario for Biosphere Catastrophe: The Permo–Triassic Boundary Case. *In* Yuen, D.A., Maruyama, S., Karato, S.-I. & Windley, B.F. (eds.), *Superplumes: Beyond Plate Tectonics*. Springer, pp. 409–440.

Mata, S.A. & Bottjer, D.J. 2011. Microbes and mass extinctions: paleoenvironmental distribution of microbialites during times of biotic crisis. *Geobiology* 10, 3–24.

McGhee, G.R., Jr., Clapham, M.E., Sheehan, P.M., Bottjer, D.J. & Droser, M.L. 2013. A new ecological-severity ranking of major Phanerozoic biodiversity crises. *Palaeogeography, Palaeoclimatology, Palaeoecology* 370, 260–270.

Payne, J.L. & Kump, L.R. 2007. Evidence for recurrent Early Triassic massive volcanism from quantitative interpretation of carbon isotope fluctuations. *Earth and Planetary Science Letters* 256, 264–277.

Pietsch, C. & Bottjer, D.J. 2014. The importance of oxygen for the disparate recovery patterns of the benthic macrofauna in the Early Triassic. *Earth-Science Reviews* 137, 65–84.

Pruss, S.B. & Bottjer, D.J. 2005. The reorganization of reef communities following the end-Permian mass extinction. *Comptes Rendus PALEVOL* 4, 553–568.

Pruss, S.B., Bottjer, D.J., Corsetti, F.A. & Baud, A. 2006. A global marine sedimentary response to the end-Permian mass extinction: Examples from southern Turkey and the western United States. *Earth-Science Reviews* 78, 193–206.

Pruss, S., Fraiser, M. & Bottjer, D.J. 2004. Proliferation of Early Triassic wrinkle structures: Implications for environmental stress following the end-Permian mass extinction. *Geology* 32, 461–464.

Pruss, S.B., Payne, J.L. & Bottjer, D.J. 2007. *Placunopsis* bioherms: The first metazoan buildups following the end-Permian mass extinction. *Palaios* 22, 17–23.

Reichow, M.K., Pringle, M.S., Al'Mukhamedov, A.I., Allen, M.B., Andreichev, V.L., Buslov, M.M., Davies, C.E., Fedoseev, G.S., Fitton, J.G., Inger, S., Medvedev, A. Ya., Mitchell, C., Puchkov, V.N., Safonova, I.Yu., Scott, R.A. & Saunders, A.D. 2009. The timing and extent of the eruption of the Siberian Traps large igneous province: Implications for the end-Permian environmental crisis. *Earth and Planetary Science Letters* 277, 9–20.

Sogot, C.E., Harper, E.M., & Taylor, P.D. 2013. Biogeographical and ecological patterns in bryozoans across the Cretaceous-Paleogene boundary: Implications for the phytoplankton collapse hypothesis. *Geology* 41, 631–634.

Song, H., Tong, J. & Chen, Z.Q. 2011. Evolutionary dynamics of the Permian–Triassic foraminifer size: Evidence for Lilliput effect in the end-Permian mass extinction and its aftermath. *Palaeogeography, Palaeoclimatology, Palaeoecology* 308, 98–110.

Stanley, S.M. 2009. Evidence from ammonoids and conodonts for multiple Early Triassic mass extinctions. *Proceedings of the National Academy of Sciences* 106, 15264–15267.

Sun, Y., Joachimski, M.M., Wignall, P.B., Yan, C., Chen, Y., Jiang, H., Wang, L. & Lai, X. 2012. Lethally hot temperatures during the Early Triassic greenhouse. *Science* 338, 366–370.

Twitchett, R.J. 2007. The Lilliput effect in the aftermath of the end-Permian extinction event. *Palaeogeography, Palaeoclimatology, Palaeoecology* 252, 132–144.

Twitchett, R.J. & Wignall, P.B. 1996. Trace fossils and the aftermath of the Permo-Triassic mass extinction: evidence from northern Italy. *Palaeogeography, Palaeoclimatology, Palaeoecology* 124, 137–151.

Twitchett, R.J., Krystyn, L., Baud, A., Wheeley, J.R. & Richoz, S. 2004. Rapid marine recovery after the end-Permian mass-extinction event in the absence of marine anoxia. *Geology* 32, 805–808.

Twitchett, R.J. 2006. The palaeoclimatology, palaeoecology and palaeoenvironmental analysis of mass extinction events. *Palaeogeography, Palaeoclimatology, Palaeoecology* 232, 190–213.

Zonneveld, J.-P., MacNaughton, R.B., Utting, J., Beatty, T.W., Pemberton, S.G. & Henderson, C.M. 2010. Sedimentology and ichnology of the Lower Triassic Montney Formation in the Pedigree-Ring/Border-Kahntah River area, northwestern Alberta and northeaster British Columbia. *Bulletin of Canadian Petroleum Geology* 58, 115–140.

Additional reading

Hart, M.B. (ed.). 1996. Biotic Recovery from Mass Extinction Events. *Geological Society Special Publications* 102. The Geological Society of London.

Nichols, D.J. & Johnson, K.R. 2008. *Plants and the K–T Boundary*. Cambridge University Press.

14 Conservation paleoecology

Introduction

Due to human activity, Earth's ecosystems have been changing and will continue to change. There is a real need to understand as best as possible how in the future ecosystems will change. This is not only due to a desire to maintain natural ecosystems but also because human civilization is dependent on ecosystem services. The record of how natural ecosystems have responded to change in the past rests with the information that can be discerned through studies of paleoecology. And there is potentially a vast storehouse of information on how ecosystems respond to the environmental changes Earth is currently experiencing.

Conservation biology is a branch of modern biology that has rapidly grown, largely through the use of modern ecological data and principles, as one way to address these needs. Conservation paleobiology encompasses all that paleontology can bring to the table for managing these crucial problems. Conservation paleoecology is part of this effort, and several conceptual and analytical approaches of this new field are outlined later.

Shifting baselines

Environmental change has become so extensive that ecosystems which reflect prehuman structure are likely found only through examination of the past. This concept has been developed through the appreciation of "shifting baselines," promoted by Jeremy Jackson and others, whereby each generation of human observers interprets what constitutes a "natural" environment to be what they observe in settings during their lifetimes. And, thus, if ecosystems are to be restored, that would be to the currently understood natural state. But if there has been change occurring over multiple generations, this baseline would be changing. Ecosystem restoration would then only result in some earlier degraded state, rather than the state occurring before human-induced environmental stress began.

The shifting baseline is illustrated in various ways in marine ecosystems that have experienced fishing. The food webs in kelp forests, coral reefs, and estuaries have changed dramatically due to extensive fishing activity (Fig. 1.8). In more specific cases, the size of fish, such as cod, has been reduced dramatically in catches over the past 1,000 years (Fig. 14.1), as have the number of oysters gathered from Chesapeake Bay (Fig. 14.1). Similarly, the percent of Caribbean reef localities with the coral *Acropora cervicornis* has declined dramatically over the past 100 years. It is only through historical data as well as that from the recent fossil record that these changes have been determined and from which knowledge has been obtained on how ecosystems looked and behaved before extensive human influence.

Paleoecology: Past, Present and Future, First Edition. David J. Bottjer.
© 2016 John Wiley & Sons, Ltd. Published 2016 by John Wiley & Sons, Ltd.

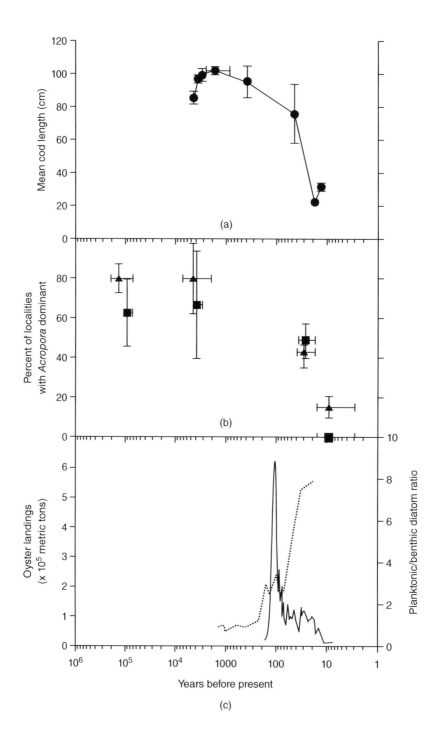

Figure 14.1 Shifting baselines in marine ecosystems. This shows retrospective data showing baselines before ecosystem collapse. (a) Time series of mean body length of Atlantic cod from kelp forests in the coastal Gulf of Maine. The first five data points are derived from archeological records, whereas the last three points are from fisheries data. Vertical bars represent the standard error. Horizontal bars represent the time range of data for a single interval of observations. (b) Paleoecological and ecological data showing the percentage of Caribbean localities with *Acropora palmata* (triangles) or *A. cervicornis* (rectangles) as the dominant shallow-water coral in the Late Pleistocene, Holocene, before 1983, and after 1983. Percentages of localities are significantly different over the four time periods for *A. palmata* and *A. cervicornis*. Vertical and horizontal bars are as in (a). (c) Paleoecological and fisheries data from Chesapeake Bay showing the ratio in abundance of planktonic to benthic diatoms (dotted line) and landings of the oyster *Crassostrea virginica* (solid line). The planktonic to benthic diatom ratio is a proxy for eutrophication that shows the relative amount of planktonic to benthic primary production. For over 1,200 years, this ratio remained fairly constant at about 1:1, but then increased threefold coincidentally with increased runoff of sediments and nutrients due to European agriculture after 1750. The ratio remained at about 3:1 between 1830 and 1930, after which it increased dramatically to about 8:1. Oyster landings show an initial increase in the early 19th century, peak in 1884, and subsequently collapse as deep channel reefs were destroyed by mechanical dredging. These data strongly imply that oysters were able to limit the potential for eutrophication induced by increased inputs of nutrients between 1750 and 1930 until oyster populations collapsed as a result of overfishing. From Jackson et al. (2001). Reproduced with permission from the American Association for the Advancement of Science.

Nonanalog communities and exotic species

As atmosphere and ocean temperatures increase, we can expect migration toward the poles of organisms that are sensitive to these temperature changes. Looking to the past to see how such changes occur will prove to be fruitful. Of course, we have a record of such changes from the most recent transition from glacial to today's interglacial times, although this record is for a change from colder to warmer environments on Earth. Fundamental observations from terrestrial environments are that many individual species commonly migrate latitudinally on an individualistic basis due to their particular environmental tolerances. This can lead to different community assemblages than what are found today, and these have been termed nonanalog assemblages that lived in novel ecosystems. This phenomenon is illustrated in Fig. 14.2 of pollen assemblages from 8,000 to 17,000 years ago, where

Figure 14.2 Development of nonanalog terrestrial communities. This is a summary pollen diagram from Appleman Lake, Indiana (United States), for the period 8,000–17,000 years ago, when there was a global transition from glacial to interglacial conditions. Only the major tree pollen types are shown. Pollen assemblages with nonanalog modern pollen assemblages occur within 13,700–11,900 years ago (gray-shaded area). The percentage of spores of the dung fungus *Sporormiella*, a record of megafaunal presence, and number of charcoal particles, a record of fire frequency and extent, are also shown. This multiproxy record suggests that the nonanalogous pollen assemblages closely followed in time the extinction of the Pleistocene megafauna, whereas enhanced fire regimes began soon after the nonanalogous assemblages. Loss of mega-herbivores may have altered ecosystem processes by the release of palatable hardwood trees from herbivore pressure and the accumulation of combustible fuel. The nonanalog pollen assemblages may thus have resulted from novel climates and release from megafaunal herbivory. From Willis et al. (2010). Reproduced with permission from Elsevier. (*See insert for color representation.*)

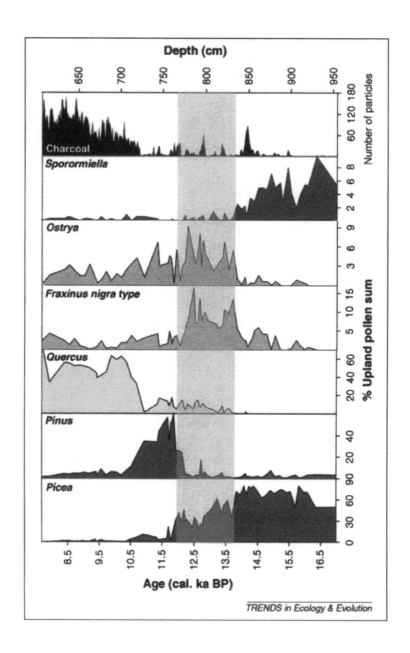

TRENDS in Ecology & Evolution

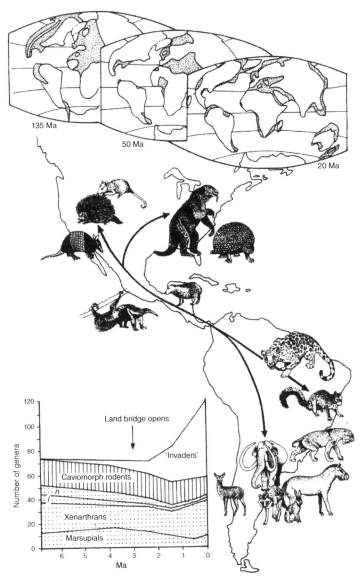

Figure 14.3 Breakdown of biogeographic barriers and migration of exotic species. The development of the Isthmus of Panama (top) promoted the great American biotic interchange between North and South American terrestrial vertebrates, which had been isolated from each other for more than 70 million years. Trends in number of genera of mammalian types in South America through the interchange (bottom) show an increase in invading placentals 2 million years ago; L, litopterns; n, notoungulates. Litopterns and notoungulates are extinct orders of hoofed mammalian ungulates; xenarthrans are a group of placental mammals including anteaters, tree sloths, and armadillos; caviomorph rodents uniquely diversified in South America. South American mammals that invaded North America (middle) include the armadillo, opossum, and porcupine, which are still common today. From Benton and Harper (2009). Reproduced with permission from John Wiley & Sons.

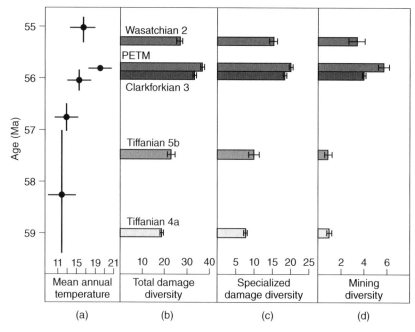

Figure 14.4 Increase in insect herbivory during an episode of global warming. This shows changes in the diversity of insect damage types on leaves of several Paleocene–Eocene floras before, during, and after the Paleocene–Eocene Thermal Maximum (PETM). Damage diversity is the percentage of the total number of possible insect damage types that could be found that are recorded from a particular sample of leaves. (a) Estimates of mean annual paleotemperature through the studied interval. (b) Total damage diversity on each flora standardized to 800 leaves, with error bars representing 1 standard deviation above and below the mean of the resamples. (c) Specialized damage diversity, which is a damage made by insects that usually consume only one or a few closely related plant species, standardized as in (b). (d) Diversity of mine morphotypes standardized as in (b); leaf miners are insect larvae that live in leaves and eat leaf tissues. From Currano et al. (2008). Reproduced with permission from the National Academy of Sciences.

pollen assemblages that do not have a modern analog occurred 12,000–13,500 years ago. That novel ecosystems may develop in the future with increased global warming is a phenomenon that we would not be able to anticipate without these paleoecological studies.

In modern ecosystems, organism migration is occurring not only due to global warming but also due to human transport of animals and plants, so-called exotic species, to new environments. The way that migrations of exotic species play out can also potentially be studied in deep time, where natural processes have initiated breakdowns of biogeographic provinces and invasions by exotic species. One of the best known examples of this phenomenon is the "Great American Faunal Interchange," which occurred with development of the Central American land bridge that tectonically formed in the Pliocene (Fig. 14.3). Prior to this time, South American mammal faunas were composed primarily of marsupials and the placental xenarthrans (Fig. 14.3), while North American mammal faunas were composed of a variety of other placentals. Migration of mammals sent North American placentals south and marsupials and xenarthrans north, with the placental invasion of

South America being more successful than that of the marsupials and xenarthrans. Future mixing of terrestrial and marine faunas will also produce winners and losers.

Ancient hyperthermal events

Going to even deeper time, intervals of sudden global warming due to CO_2 injection into the atmosphere by natural processes, termed hyperthermals, are worthy of significant study. At the boundary between the Paleocene and the Eocene, there is a geologically sudden temperature rise (Fig. 14.4), the so-called Paleocene–Eocene Thermal Maximum (PETM). At this time, there is documented migration to the poles of terrestrial assemblages as well as a decrease of size in mammalian lineages. Intriguingly, studies of fossil leaves from before, during, and after this interval show an increase in vegetation destruction due to insect activity (Fig. 14.4). This is a particularly fascinating piece of paleoecological data when one is considering management of forest ecosystems and agricultural settings in the future as the Earth warms. The PETM has been used as a model for predicting the outcomes of current global warming. However, it has been suggested that the PETM lacked the severity of the current rate of change, and thus, it becomes crucial that even more severe hyperthermal events in Earth's past be examined.

As discussed in Chapter 13, the most severe biotic crisis since the Phanerozoic was due to injection of carbon dioxide and other greenhouse gases into the atmosphere by eruption of the Siberian Traps at the end of the Permian (Fig. 13.1). During this time, the Earth's continents were assembled into one giant supercontinent, Pangea, which would have affected climate differently than today's dispersed continents. Plankton with calcium carbonate skeletons had not yet evolved, so the extent, and hence calcium carbonate buffering capacity, of seafloors covered in carbonate sediments was more restricted than in modern oceans (Fig. 12.18). Biologically, the organisms that suffered the greatest extinction

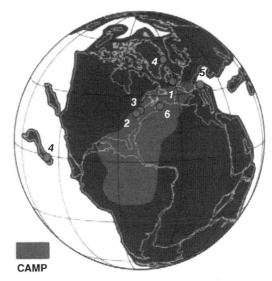

CAMP

Figure 14.5 Pangea and the areal distribution of CAMP volcanism. This also shows the location of major localities where CAMP and the Triassic–Jurassic mass extinction have been studied. (1) St. Audries Bay; (2) Newark basin; (3) Hartford basin; (4) Kennecott Point; (5) Val Adrara; and (6) Moroccan CAMP sections. From Whiteside et al. (2010). Reproduced with permission from the National Academy of Sciences. (*See insert for color representation.*)

were members of the Paleozoic marine evolutionary fauna, which is much diminished in today's oceans (Figs. 3.11 and 14.6). Thus, paleogeographic, climatic, geochemical, and biological conditions on Earth were different then than now. Despite these differences, examination of the paleoecology of the end-Permian mass extinction interval is highly warranted to understand the ecological course of our future global warming ocean.

To summarize from Chapter 13, the ocean then was characterized by widespread oxygen-minimum zones, which impinged from deep into shallow environments, and typically were euxinic (Fig. 13.11). The effects of this mass extinction on land led to a significant increase of terrigenous runoff to shallow-marine systems. The prevalence of microbial structures implies an ocean more dominated by microbes than is typical of the Phanerozoic (Figs. 13.4 and 13.5). These stressful environmental

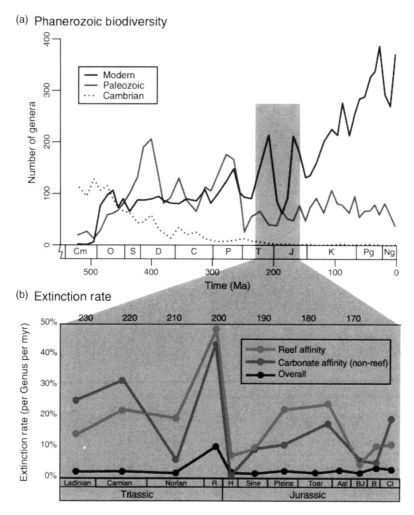

Figure 14.6 The end-Triassic extinction and extinction severity for organisms with an affinity for carbonate environments. (a) Phanerozoic biodiversity, note that the end-Triassic extinction represents perhaps the most severe Phanerozoic biodiversity crisis for the Modern marine evolutionary fauna. (b) Extinction rates through the Middle and Late Triassic to the Early Middle Jurassic. Extinction rates of reef genera and level bottom (nonreef) genera that have an affinity for carbonate substrates are compared with the overall extinction rates of benthic invertebrate taxa. Note how Rhaetian (R) extinction rates are much higher for both carbonate level bottom faunas and reef faunas than the overall extinction rate and background levels. Extinction rates are binned by stage and therefore plotted at the midpoint of each age interval. From Greene et al. (2012). Reproduced with permission from Elsevier. (*See insert for color representation.*)

conditions, including acidification, also led to an Early Triassic metazoan reef gap followed by a reef eclipse interval (Figs. 13.7–13.9). Level bottom benthic environments were characterized by low biodiversity, cosmopolitan taxa, and reduced depth and extent of bioturbation (Figs. 1.3, 13.12, 13.16, and 13.17). The extreme heat of much of the Early Triassic led organisms to shift their distributions

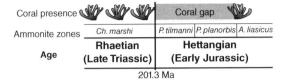

Coral presence				Coral gap		
Ammonite zones	*Ch. marshi*		*P. tilmanni*	*P. planorbis*	*A. liasicus*
Age	**Rhaetian** (Late Triassic)		**Hettangian** (Early Jurassic)		
		201.3 Ma			

Figure 14.7 Duration of coral and coral reef gap after the end-Triassic mass extinction. The extent of this coral gap is ~370,000 years. From Martindale et al. (2012). Reproduced with permission from Elsevier.

poleward (Figs. 13.15 and 13.21). Many of these ecological conditions that occurred following the end-Permian mass extinction bear striking similarity to what current predictions envisage for ocean ecosystems in the future. Thus, further detailed study of the effects of this mass extinction should help provide guidelines to manage future ecological changes.

Another hyperthermal event in deep time warranting extensive study is the end-Triassic mass extinction, which is the most recent of the Big Five mass extinctions to have primarily been caused by carbon dioxide injection into the atmosphere. This carbon dioxide injection was the result of eruption of the Central Atlantic Magmatic Province (CAMP) (Fig. 14.5), which initiated the rifting of Pangea that formed the central Atlantic Ocean. Paleogeographic, climatic, and geochemical conditions on Earth were similar to those during the end-Permian mass extinction. However, in the oceans, the end-Triassic mass extinction was the biggest mass extinction for the Modern marine evolutionary fauna (Fig. 14.6) and thus a more direct biological analog for today's marine organisms.

Scleractinian corals first appear in the stratigraphic record in the Middle Triassic (Fig. 13.7). These corals are commonly the dominant reef-building organisms from then until now except for the interval of rudist dominance in the Cretaceous (Fig. 12.17). In particular, the Late Triassic was a time of spectacular reef development, and scleractinian corals played a prominent role in this (Figs. 9.8 and 9.9). However, beginning at the Triassic–Jurassic boundary, there is a reef and

coral gap that lasted several hundred thousand years (Fig. 14.7). This absence of corals as part of the mass extinction may have been due to episodes of ocean acidification due to CAMP eruptions. Ocean acidification as a significant process causing this mass extinction is also indicated because extinction rates for organisms with a reef affinity or a nonreef carbonate affinity (Fig. 14.6) are much higher than for the overall marine fauna. Data from Triassic–Jurassic boundary sections reinforce the interpretation that this mass extinction affected the carbonate chemistry of the oceans, as many of these sections show a significant reduction in carbonate production across this interval (Fig. 14.8). Since ocean acidification is an ongoing problem causing distress to modern marine ecosystems, this mass extinction shows unusual promise for understanding the effects that such changes have in marine environments and what we might expect for our future global acidifying ocean.

Bioturbation in marine sediments was also affected (Fig. 14.9), although not to the extent seen after the end-Permian mass extinction. Dramatic changes in abundance and dominance of organic-walled green algal phytoplankton are also recorded after this mass extinction (Fig. 14.10), interpreted as disaster species and reflecting the stressed nature of the ecosystem.

Summary

Study of the end-Cretaceous impact mass extinction has led to increased awareness of the possibility of modern asteroid impacts. In the same way, intensive study of early Mesozoic hyperthermals that led to mass extinctions has much to teach us about the ecology of the future greenhouse world (Fig. 1.9). Thus, paleoecological studies of past environmental change show that, during periods of geologically rapid environmental stress, what seem like coherent communities of organisms can disassemble as individual species either become extinct or seek new locations that more adequately fit their range of environmental sensitivities. These geologically

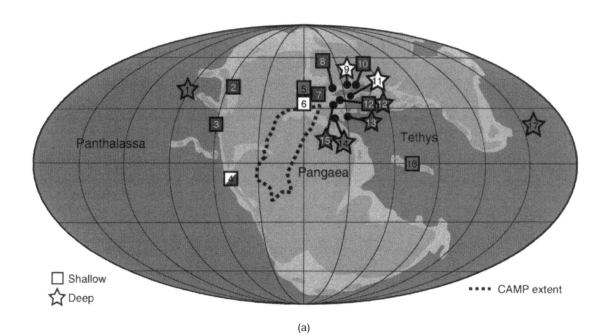

(a)

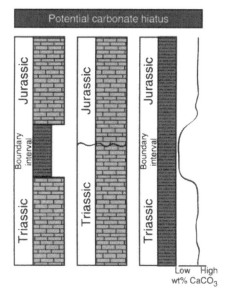

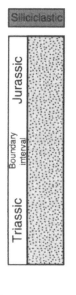

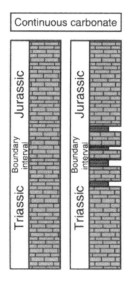

(b)

Figure 14.8 Reduction of carbonate deposition during the end-Triassic mass extinction. This is illustrated through examination of global lithological changes in Triassic–Jurassic boundary sections. (a) Early Jurassic paleogeographic reconstruction with the hypothesized extent of CAMP flood basalts (dashed outline) and the approximate paleolocation of Triassic–Jurassic boundary sites plotted. Squares represent shallow-water (shelf) sections, while stars represent deepwater (basinal) sections. Black indicates a potential carbonate hiatus across the boundary, dark gray indicates predominantly siliciclastic sections, and white indicates either a section with potentially continuous carbonate deposition across the boundary or a section without reliable microstratigraphic data. (b) Idealized stratigraphic columns of each section type (carbonate hiatus, siliciclastic, and potentially continuous carbonate). 1, Queen Charlotte Islands, British Columbia, Canada; 2, Williston Lake, British Columbia, Canada; 3, New York Canyon, Nevada, United States; 4, Chilingote, Utcubamba Valley, Peru; 5, St. Audries Bay, England; 6, Asturias, northern Spain; 7, Mingolsheim core, Germany; 8, Northern Calcareous Alps, Austria; 9, Csovar section, northern Hungary; 10, Tatra Mountains (Carpathians), Slovakia; 11, Tolmin Basin (Southern Alps), Slovenia; 12, Lombardian Basin (Southern Alps), Italy; 13, Budva Basin (Dinarides), Montenegro; 14, Southern Apennines, Italy; 15, Northern and Central Apennines, Italy; 16, Germig, Tibet; 17, Southwest Japan. From Greene et al. (2012). Reproduced with permission from Elsevier. (*See insert for color representation.*)

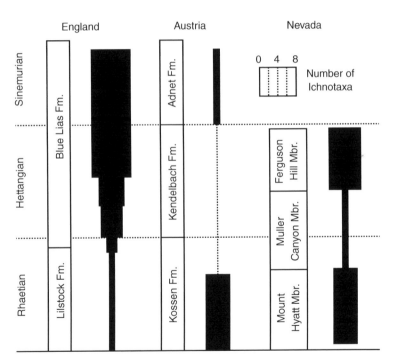

Figure 14.9 Trends in ichnodiversity through the Triassic–Jurassic intervals of England, Central Austria, and Nevada, United States. Additional studies show evidence for reduced burrow size, depth of bioturbation, and record of ichnofabric indices in the uppermost Rhaetian (latest Triassic) and an increase in these parameters through the Hettangian (earliest Jurassic). Studied section in England is at St. Audries Bay and Pinhay Bay; in Austria at Kendelbach Gorge and a nearby area; in Nevada at the New York Canyon area of the Gabbs Valley Range. Gap in the Austrian record is the result of diagenesis. From Twitchett and Barras (2004). Reproduced with permission from the Geological Society.

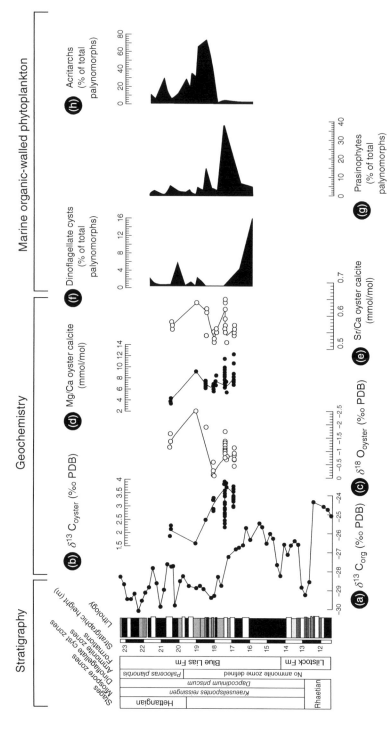

Figure 14.10 Early Jurassic phytoplankton disaster species at St. Audries Bay (England). This illustrates phytoplankton abundance and geochemical trends in this Triassic–Jurassic stratigraphic section. (a) Stratigraphy, lithology, and carbon isotope record for bulk organic matter. (b) Carbon isotope record for oysters. (c) Oxygen isotope record for oysters. (d) Mg/Ca record from oyster calcite. (e) Sr/Ca record from oyster calcite. (f) Dinoflagellate cyst abundance as a percentage of the total palynomorph fraction. (g) Prasinophyte phycomata abundance as a percentage of the total palynomorph fraction. (h) Acritarch abundance as a percentage of the total palynomorph fraction. Percentage plots show "blooms" in abundance of prasinophytes (40%) and acritarchs (80%) near the first occurrence of the earliest Early Jurassic ammonite *Psiloceras planorbis*. The geochemical plots from analyses of samples from shells of the oyster *Liostrea hisingeri* have 20% weighted average curve fits to highlight changes in the various proxies. An "initial" negative organic carbon isotope excursion at and adjacent to bed 13 has been considered to represent the Triassic–Jurassic boundary. A second "main" carbon isotope excursion above bed 17 measured from organic carbon (a) is replicated in oyster carbonate carbon measurements (b). Oyster oxygen isotope, Mg/Ca, and Sr/Ca data all indicate an increase in water temperatures during these phytoplankton "blooms" and this part of the "main" carbon isotope excursion. From van de Schootbrugge et al. (2007). Reproduced with permission from Elsevier.

rapid changes in environmental conditions causing extinction resulting in open and new ecospace then also provide opportunities for organisms well adapted to the new conditions to become abundant and to diversify. These ecosystem changes can also lead to reductions in size of component organisms and an increase in microbial structuring of environments in sedimentary settings. Therefore, growing development of a deep time perspective for conservation paleoecology provides many new research directions for paleoecological studies. As the present has been the key to the past, the past can be the key to the future.

References

Benton, M.J. & Harper, D.A.T. 2009. *Introduction to Paleobiology and the Fossil Record*. Wiley-Blackwell.

Currano, E.D., Wilf, P., Wing, S.L., Labandeira, C.C., Lovelock, E.C. & Royer, D.L. 2008. Sharply increased insect herbivory during the Paleocene–Eocene Thermal Maximum. *Proceedings of the National Academy of Sciences* 105, 1960–1964.

Greene, S.E., Martindale, R.C., Ritterbush, K.A., Bottjer, D.J., Corsetti, F.A. & Berelson, W.M. 2012. Recognising ocean acidification in deep time: An evaluation of the evidence for acidification across the Triassic–Jurassic boundary. *Earth-Science Reviews* 113, 72–93.

Jackson, J.B.C., Kirby, M.X., Berger, W.H., Bjorndal, K.A., Botsford, L.W., Bourque, B.J., Bradbury, R.H., Cooke, R., Erlandson, J., Estes, J.A., Hughes, T.P., Kidwell, S., Lange, C.B., Lenihan, H.S., Pandolfi, J.M., Peterson, C.H., Steneck, R.S., Tegner, M.J. & Warner, R.R. 2001. Historical overfishing and the recent collapse of coastal ecosystems. *Science* 293, 629–637.

Martindale, R.C., Berelson, W.M., Corsetti, F.A., Bottjer, D.J. & West, A.J. 2012. Constraining carbonate chemistry at a potential ocean acidification event (the Triassic–Jurassic boundary) using the presence of corals and coral reefs in the fossil record. *Palaeogeography, Palaeoclimatology, Palaeoecology* 350–352, 114–123.

van de Schootbrugge, B., Tremolada, F., Rosenthal, Y., Bailey, T.R., Feist-Burkhardt, S., Brinkjuis, H., Pross, J., Kent, D.V. & Falkowski, P.G. 2007. End-Triassic calcification crisis and blooms of organic-walled disaster species. *Paleogeography, Palaeoclimatology, Palaeoecology* 244, 126–141.

Twitchett, R.J. & Barras, C.G. 2004. *In* McIlroy, D. (ed.), *The Application of Ichnology to Palaeoenvironmental and Stratigraphic Analysis. Geological Society Special Publication* 228, 397–418.

Whiteside, J.H., Olsen, P.E., Eglinton, T., Brookfield, M.E. & Sambrotto, R.N. 2010. Compound-specific carbon isotopes from Earth's largest flood basalt eruptions directly linked to the end-Triassic mass extinction. *Proceedings of the National Academy of Sciences* 107, 6721–6725.

Willis, K.J., Bailey, R.M., Bhagwat, S.A. & Birks, H.J.B. 2010. Biodiversity baselines, thresholds and resilience: testing predictions and assumptions using palaeoecological data. *Trends in Ecology and Evolution* 25, 583–591.

Additional reading

Dietl, R.G. & Flessa, K.W. (eds.). 2009. Conservation Paleobiology: Using the Past to Manage for the Future. *The Paleontological Papers, Volume 15*. The Paleontological Society.

Louys, J. (ed.). 2012. *Paleontology in Ecology and Conservation*. Springer, Heidelberg.

Corsetti, F.A., Ritterbush, K.A., Bottjer, D.J., Greene, S.E., Ibarra, Y., Yager, J.A., West, A.J., Berelson, W.M., Rosas, S., Becker, T.W., Levine, N.M., Loyd, S.J., Martindale, R.C., Petryshyn, V.A., Carroll, N.R., Petsios, E., Piazza, O., Pietsch, C., Stellmann, J.L., Thompson, J.R., Washington, K. & Wilmeth, D.T. 2015. Investigating the paleoecological consequences of supercontinent breakup: sponges clean up in the Early Jurassic. *The Sedimentary Record* 13, 4–9.

Index

Printed and bound by CPI Group (UK) Ltd, Croydon, CR0 4YY